Ben Stacy Jerrik (Ed.)

Pete French Round Barn

Ben Stacy Jerrik (Ed.)

Pete French Round Barn

National Register of Historic Places, Burns, Oregon, Harney County, Oregon

Part Press

Imprint

Publisher:
Part Press is a trademark of
International Book Market Service Ltd., 17 Rue Meldrum, Beau Bassin, 1713-01 Mauritius
Email: info@bookmarketservice.com
Website: www.bookmarketservice.com

Published in 2013

Printed in: U.S.A., U.K., Germany. This book was not produced in Mauritius.

ISBN: 978-613-9-09923-8

Contents

Articles

References

Peter_French

Peter French	

Born	April 30, 1849Missouri, United States
Died	December 26, 1897 (aged 48)Frenchglen, Oregon, USA
Occupation	Rancher

Peter French (April 30, 1849 – December 26, 1897) was a rancher in the western United States in the late 19th century. The community of Frenchglen, Oregon was partially named for him.

Early life

Peter French was born John William French in Missouri on April 30, 1849. In 1850, his father moved the family to Colusa County, California to begin a small ranch. Finding there was not enough room for small ranch operations due to Spanish land grants, French's father uprooted his family once again and traveled north to the Sacramento Valley. French's father began a sheep ranch there, which became very successful; however, as French grew older, he found that the work was not exciting or challenging enough for him.

French moved southward to Jacinto, California where he met and accepted employment as a horse breaker with Dr. Hugh James Glenn, a wealthy stockman and wheat baron. French was a quick learner and good worker, and in a few months he was promoted to foreman. The Spanish-speaking *vaqueros* liked and respected French, as he learned their language. At some point in his employment with Glenn, French assumed the name "Peter".

Glenn had expanded his assets as widely as possible in the area, and began to scout new areas for his profitable markets. In 1872, he sent French to Oregon with 1,200 head of Shorthorn cattle, a handful of *vaqueros*, and a Chinese cook. He ended up in southeastern Oregon to find vast grasslands amid the arid high desert.

Upon his arrival in the Catlow Valley, French and his men came upon a poor prospector named Porter. Porter sold his small herd of cattle to French, and with the sale of his cattle went his squatter's rights to the west side of the Steens Mountain and his "P" brand. As French ventured further, he found the Blitzen Valley, where the Donner und Blitzen River snaked northward 40 miles to Malheur Lake. This became his favorite spot, where he set up his base camp. He built shelters for his herd, line cabins, and bunkhouses for his men. Thus, the P Ranch was born.

Cattle king

After several years, French's small cattle operation had expanded, helped in large part by Glenn as his financier. The P Ranch became the headquarters for his growing cattle empire. He and his men built fences, drained marshlands and irrigated large areas of land, broke hundreds of horses and mules, and cut and stacked native hay. French's empire expanded to include the Diamond Valley, the Blitzen Valley, and the Catlow Valley. The land encompasses approximately 160,000 to 200,000 acres (650 to 800 km²). A shrewd businessman, French took advantage of the

Swamp and Overflow Act, which allowed marshland to be purchased at $1.25 an acre. He built dams to flood areas, bought the land under the Swamp Act at the reduced price, then removed the dams to return the land to its original state. French also directed his employees and others to file homestead claims that he would then purchase. His attempts at seizing more and more land even included fencing lands in the public domain.

In 1883, French married Glenn's daughter Ella. Glenn was murdered three weeks later by a former employee. French continued to manage the Oregon operation for the Glenn family, selling more cattle to help pay the family's debts. In 1894, Glenn's heirs decided to incorporate the French-Glenn partnership into the *French-Glenn Livestock Company*, making French the company president.[1] [2] French was divorced in 1891.

Fall

In June 1878 the native Paiute and Bannock (both closely associated with the Shoshone tribes) population at the base of the Steens Mountain swooped upon the P Ranch, but not before a messenger could warn French of the impending attack. French and all but one of his men were able to escape. The attacks continued throughout the summer. The Paiutes burned buildings and homes, ran off cattle and horses, and at least once shot French's horse out from under him. At one point, French even accompanied the U.S. 1st Cavalry Regiment to guide the Army through the area.

In the 1880s and 1890s, stockmen and smaller farmers fought over land and water rights. Aggression over such rights and French's large spread of land drew a certain loathing toward him.

French arrived back home from a business trip on Christmas Day in 1897 bearing gifts he had purchased for his wife, children and cowhands. His crew chief fell ill in the night, and French took on the responsibilities of trail boss the next morning.

On December 26, 1897, French opened the gate to a sagebrush field to let his cattle through. As he remounted his horse, Ed Oliver, a man with whom French had had a previous boundary spat, rode his horse at full speed straight at French. Oliver's horse collided with French's, and as Oliver charged again, French struck him over the head with a willow whip. Oliver produced a revolver from his waistband and fired one shot. The bullet struck French in the head and killed him instantly. Oliver, fearing retribution from French's men, sped off.

Another lost battle

Harney County Sheriff Andrew John McKinnon arrested Oliver the next day, Oliver still wearing the weapon that had taken French's life. French's brother, Burt, filed an indictment for murder; the charge was dropped the day before Oliver's trial was to begin. Harney County District Attorney Charles W. Parfish would not let it drop, and filed a new indictment for manslaughter, acting on the belief that he would stand a better chance of pinning down the lesser charge.

The trial began on May 19, 1898, with members of French's crew testifying that Oliver had shot French from behind as he rode away from Oliver. Lionel R. Webster, attorney for Oliver, maintained that Oliver had shot French in self-defense, having feared for his life. Grasping at straws, Webster put his client on the stand. Oliver testified that French beat him with the willow whip, and screamed "I'll kill you!" as he reached for a gun. French's men had earlier testified, however, that French was not armed, and jurors took only three hours in deliberation before returning the verdict "not guilty".

John William "Peter" French was buried in Red Bluff, California next to the graves of his father and mother at the Oak Hill Cemetery.

See also

- Pete French Round Barn

References

[1] "A Little Bit of Malheur History" (http://www.fws.gov/malheur/history.html), Malheur National Wildlife Refuge, United States Fish and Wildlife Service, United States Department of Interior, Princeton, Oregon, 10 November 2008.

[2] Pinyard, David and Donald Peting, "Preservation of the Pete French Round Barn" (http://crm.cr.nps.gov/archive/18-5/18-5-11.pdf), *CRM Cultural Resources Management* (Vol. 18, No. 5), National Park Service, United States Department of Interior, Washington, D.C., 1995, pp. 30–32.

Further reading

- French, Giles. "Cattle Country of Peter French." Binfords & Mort, 1972.
- Gibson, Elizabeth. "Pete French, Cattle Baron." http://www.suite101.com/article.cfm/old_west/97092/1
- Harney County Sherriff. "The Death of Peter French." (http://gesswhoto.com/sheriff-harney4.html)

External links

- Peter French biography (http://www.ohs.org/education/oregonhistory/Oregon-Biographies-Peter-French.cfm) from the Oregon Historical Society

Pete_French_Round_Barn

Pete French Round Barn	
U.S. National Register of Historic Places	
Location:	Harney County, Oregon
Nearest city:	Burns
Coordinates:	43°7′57″N 118°38′32″W
Area:	5 acres (ha)
Built:	c. 1875-1885[1]
Architectural style:	Round barn
Governing body:	Private
NRHP Reference#:	71000679[2]
Added to NRHP:	September 10, 1971

The **Pete French Round Barn**, located near Burns, Oregon, United States, is a round barn listed on the National Register of Historic Places. The late 19th century barn was owned and constructed by cattle rancher Peter French; French trained horses there during the winter. The barn was listed on the National Register on September 10, 1971.

History

This round barn was constructed sometime in the late 1870s or early 1880s. There were originally two—or possibly three—in the vicinity, but this is the only round barn remaining. They were used to break horses during the winter. Three hundred to one thousand horses were moved through each year.[3] The French family donated the site to the state of Oregon in 1969.[1]

Design

Interior ceiling of Pete French Round Barn shows the large single juniper center post and dimentonal wood lumber support structure

The interior of the barn consists of a circular rock wall, constructed with local rock and mud, which encloses the central stable area. The remainder of the structure is constructed from lumber that was hauled to the site from 60 miles to the north. The interior is unique in that it is supported by juniper posts. The center area was used as a horse stable; around the outside of the rock wall, there is a covered circular track that was used to exercise horses during winter months. The barn looks much as it did when cattleman Pete French constructed it.[1]

References

[1] "Round Barn Visitors Center" (http://www.roundbarn.net/roundbarn.html). Jenkins Historical Tours. . Retrieved 2007-02-08.

[2] "National Register Information System" (http://nrhp.focus.nps.gov/natreg/docs/All_Data.html). *National Register of Historic Places*. National Park Service. 2010-07-09. .

[3] "Southeastern Oregon Historical Parks brochure" (http://www.oregonstateparks.org/images/pdf/se_or_historical.pdf) (PDF). Oregon Parks and Recreation Department. June 2004. . Retrieved 2007-02-21.

National_Register_of_Historic_Places

The **National Register of Historic Places** (**NRHP**) is the United States federal government's official list of districts, sites, buildings, structures, and objects deemed worthy of preservation. A property listed in the National Register, or located within a National Register Historic District, may qualify for tax incentives derived from the total value of expenses incurred preserving the property.

Old Slater Mill, a historic district in Pawtucket, Rhode Island, was the first property listed in the National Register on November 13, 1966.[1]

The passage of the National Historic Preservation Act (NHPA) in 1966 established the National Register and the process for adding properties to it. Of the more than one million properties on the National Register, 80,000 are listed individually. The remainder are contributing resources within historic districts. Each year approximately 30,000 properties are added to the National Register as part of districts or by individual listings.

The Loren Andrus Octagon House in Washington, Michigan has been on the NRHP since September 3, 1971.

For most of its history the National Register has been administered by the National Park Service (NPS), an agency within the United States Department of the Interior. Its goals are to help property owners and interest groups, such as the National Trust for Historic Preservation, coordinate, identify, and protect historic sites in the United States. While National Register listings are mostly symbolic, their recognition of significance provides some financial incentive to owners of listed properties. Protection of the property is not guaranteed. During the nomination process, the property is evaluated in terms of the four criteria for inclusion on the National Register of Historic Places. The application of those criteria has been the subject of criticism by academics of history and preservation, as well as the public and politicians.

Chestnut Street District, in Salem, Massachusetts, USA, is operated as a historic house museum by Historic New England, is open for public tours, and is on the National Register of Historic Places.

Occasionally historic sites outside the country proper, but associated with the United States (such as the American Embassy in Tangiers) are also listed. Properties can be nominated in a variety of forms, including individual properties, historic districts, and multiple property submissions (MPS). The Register categorizes general listings into one of five types of properties: district, site, structure, building, or object. **National Register Historic Districts** are defined geographical areas consisting of contributing and non-contributing properties. Some properties are added automatically to the National Register when they become administered by the National Park Service. These include National Historic Landmarks (NHL), National Historic Sites (NHS), National Historical Parks, National Military Parks/Battlefields, National Memorials, and some National Monuments. (Federal properties can be proclaimed National Monuments by the Antiquities Act because of either their historical or natural significance. They are managed by multiple agencies. Only monuments that are historic in character and managed by the National Park Service are listed administratively in the National Register.)

Some examples are Fallingwater, Robie House, Martin Luther King Jr's Grave, and Old Slater Mill.

History

On October 15, 1966 the Historic Preservation Act created the National Register of Historic Places and the corresponding State Historic Preservation Offices (SHPO).[3] Initially, the National Register consisted of the National Historic Landmarks designated before the Register's creation, as well as any other historic sites in the National Park system.[4] Approval of the act, which was amended in 1980 and 1992, represented the first time the United States had a broad-based historic preservation policy.[3] [5] The 1966 act required those agencies to work in conjunction with the SHPO and an independent federal agency, the Advisory Council on Historic Preservation (ACHP), to confront adverse effects of federal activities on historic preservation.[6]

George B. Hartzog, Jr. Director of the National Park Service from January 8, 1964 until December 31, 1972 [2]

To administer the newly created National Register of Historic Places, the National Park Service of the U.S. Department of Interior, with director George B. Hartzog, Jr., established an administrative division named the Office of Archeology and Historic Preservation (OAHP).[6] [7] Hartzog charged OAHP with creating the National Register program mandated by the 1966 law. Ernest Connally was the Office's first director. Within OAHP new divisions were created to deal with the National Register.[8] The division administered several existing programs, including the Historic Sites Survey and the Historic American Buildings Survey, as well as the new National Register and Historic Preservation Fund.[6]

U.S. Secretary of the Interior (1977–1981) Cecil D. Andrus removed the National Register from the jurisdiction of the National Park Service in 1978

The first official Keeper of the Register was William J. Murtagh, an architectural historian.[4] During the Register's earliest years in the late 1960s and early 1970s, organization was lax and SHPOs were small, understaffed, and underfunded.[7] However, funds were still being supplied for the Historic Preservation Fund to provide matching grants-in-aid to listed property owners, first for house museums and institutional buildings, but later for commercial structures as well.[6]

A few years later in 1979, the NPS history programs affiliated with both the U.S. National Parks system and the National Register were categorized formally into two "Assistant Directorates." Established were the Assistant Directorate for Archeology and Historic Preservation and the Assistant Directorate for Park Historic Preservation.[8] From 1978 until 1981, the main agency for the National Register was the Heritage Conservation and Recreation Service (HCRS) of the United States Department of Interior.[9]

In February 1983, the two assistant directorates were merged to promote efficiency and recognize the interdependency of their programs. Jerry L. Rogers was selected to direct this newly merged associate directorate. He was described as a skilled administrator, who was sensitive to the need for the NPS to work with SHPOs, academia, and local governments.[8]

Although not described in detail in the 1966 act, SHPOs eventually became integral to the process of listing properties on the National Register. The 1980 amendments of the 1966 law further defined the responsibilities of SHPOs concerning the National Register.[9] Several 1992 amendments of the NHPA added a category to the National Register, known as Traditional Cultural Properties: those properties associated with Native American or Hawaiian groups.[5]

The National Register of Historic Places has grown considerably from its legislative origins in 1966. In 1986 citizens and groups nominated 3,623 separate properties, sites, and districts for inclusion on the National Register, a total of 75,000 separate properties.[9] Of the more than one million properties on the National Register, 80,000 are listed individually. Others are listed as contributing members within historic districts.[6] [10]

Property owner incentives

Properties are not protected in any strict sense by the Federal listing. States and local zoning bodies may or may not choose to protect listed historic places. Indirect protection is possible, by state and local regulations on development of National Register properties, and by tax incentives.[11]

Mohonk Mountain House, a resort hotel located on the Shawangunk Ridge

Until 1976, federal tax incentives were virtually non-existent for buildings on the National Register. Before 1976 the federal tax code favored new construction rather than the reuse of existing, sometimes historical, structures.[6] In 1976, the tax code was altered to provide tax incentives that promote preservation of income-producing historic properties. The National Park Service was given the responsibility to ensure that only rehabilitations that preserved the historic character of a building would qualify for federal tax incentives. A qualifying rehabilitation is one that the NPS deems consistent with the Secretary of the Interior's Standards for Rehabilitation.[12] Properties and sites listed in the Register, as well as those located in and contributing to the period of significance of National Register Historic Districts, became eligible for the federal tax benefits.[6]

Owners of income-producing properties listed individually in the National Register of Historic Places or of properties that are contributing resources within a National Register Historic District may be eligible for a 20% investment tax credit for the rehabilitation of the historic structure. The rehabilitation may be of a commercial, industrial, or residential property, for rentals.[13] The tax incentives program is operated by the Federal Historic Preservation Tax Incentives program, which is managed jointly by the National Park Service, individual State Historic Preservation Offices, and the Internal Revenue Service.[14] Aside from the 20% tax credit, the tax incentive program offers a 10% tax credit for rehabilitation to owners of non-historic, non-residential buildings constructed before 1936.[15]

Some property owners may qualify for grants as well, for instance the Save America's Treasures grants, which apply specifically to properties entered in the Register with national significance or designated as National Historic Landmarks.[16]

The NHPA did not distinguish between properties listed in the National Register of Historic Places and those designated as National Historic Landmarks concerning qualification for tax incentives or grants. This was deliberate, as the authors of the act had learned from experience that distinguishing between categories of significance for such incentives caused the lowest category to become expendable.[4] Essentially, this made the Landmarks a kind of "honor roll" of the most significant properties of the National Register of Historic Places.[4]

Nomination process

> It is hereby declared to be the policy of the United States Government that special effort should be made to preserve the natural beauty of the countryside and public park and recreation lands, wildlife and waterfowl refuges, and historic sites.[17]
>
> — (49 USC 303)

Any individual can prepare a National Register nomination, although historians and historic preservation consultants often are employed for this work. The nomination consists of a standard nomination form and contains basic information about a property's physical appearance and the type of significance embodied in the building, structure, object, site, or district.[18] The State Historic Preservation Office receives National Register nominations and provides feedback to the nominating individual or group. After preliminary review, the SHPO sends each nomination to the state's historic review commission, which then recommends whether the State Historic Preservation Officer should send the nomination to the Keeper of the National Register. For any non-Federally owned property, only the State Historic Preservation Officer may officially nominate a property for inclusion in the National Register. After the nomination is recommended for listing in the National Register by the SHPO, the nomination is sent to the National Park Service, which approves or denies the nomination. If approved, the property is entered officially by the Keeper of the National Register into the National Register of Historic Places.[18] Property owners are notified of the nomination during the review by the SHPO and state's historic review commission. If an owner objects to a nomination of private property, or in the case of a historic district, a majority of owners, then the property cannot be listed in the National Register of Historic Places.[18]

Criteria

For a property to be eligible for the National Register, it must meet at least one of the four National Register main criteria.[20] Information about architectural styles, association with various aspects of social history and commerce, and ownership are all integral parts of the nomination. Each nomination contains a narrative section that provides a detailed physical description of the property and justifies why it is significant historically with regard either to local, state, or national history. The four National Register of Historic Places criteria are the following.

The Robie House, designed by Frank Lloyd Wright, is an example of a property listed by means of criterion C.[19]

- **Criterion A**, "Event," the property must make a contribution to the major pattern of American history.
- **Criterion B**, "Person," is associated with significant people of the American past.
- **Criterion C**, "Design/Construction," concerns the distinctive characteristics of the building by its architecture and construction, including having great artistic value or being the work of a master.
- **Criterion D**, "Information potential," is satisfied if the property has yielded or may be likely to yield information important to prehistory or history.[18]

The criteria are applied differently for different types of properties; for instance, maritime properties have application guidelines different from those of buildings.[20]

Exclusions

There are specific instances where properties usually do not merit listing in the National Register. As a general rule, cemeteries, birthplaces, graves of historical figures, properties owned by religious institutions or used for religious purposes, moved structures, reconstructed historic buildings, commemorative properties, and properties that have achieved significance during the last fifty years are not qualified for listing on the Register.[18] There are, however, exceptions to all the preceding; mitigating circumstances allow properties classified in one of those groups to be included.[18]

Properties listed

A listing on the National Register of Historic Places is governmental acknowledgment of a historic district, site, building, or property. However, the Register is mostly "an honorary status with some federal financial incentives."[21] The National Register of Historic Places automatically includes all National Historic Landmarks as well as all historic areas administered by the National Park Service.[3] Landmarks such as these include: National Historic Sites (NHS), National Historical Parks, National Military Parks/Battlefields, National Memorials, and some National Monuments. Occasionally historic sites outside the country's borders, but associated with the United States, such as the American Embassy in Tangiers, also are listed.[22]

A typical plaque found on properties listed in the National Register of Historic Places

Listing in the National Register does not restrict private property owners from the use of their property.[13] Some states and municipalities, however, may have laws that become effective when a property is listed in the National Register. If federal money or a federal permitting process is involved, Section 106 of the National Historic Preservation Act of 1966 is invoked. Section 106 requires the federal agency involved to assess the effect of its actions on historic resources.[3] Statutorily, the Advisory Council on Historic Preservation (ACHP) has the most significant role by Section 106 of the National Historic Preservation Act. The section requires that the director of any federal agency with direct or indirect jurisdiction of a project that may affect a property listed in the National Register of Historic Places, must first report to the Advisory Council. The director of said agency is required to "take into account the effect of the undertaking" on the National Register property, as well as to afford the ACHP a reasonable opportunity to comment.[23]

The tomb of Martin Luther King Jr. in Atlanta, Georgia is part of a National Historic Site.

While Section 106 does not mandate explicitly that any federal agency director accept the advice of the ACHP, their advice has practical influence, especially given the statutory obligations of the NHPA that require federal agencies to "take into account the effect of the undertaking."[3] [23]

In cases where the ACHP determines federal action will have an "adverse effect" on historic properties, mitigation is sought. Typically, a Memorandum of Agreement (MOA) is created by which the involved parties agree to a particular plan. Many states have laws similar to Section 106.[24] In contrast to conditions relating to a federally designated historic district, municipal ordinances governing local historic districts often restrict certain kinds of changes to properties. Thus they may protect the property more than a National Register listing does.[25]

The Department of Transportation Act, passed on October 15, 1966, the same day as the National Historic Preservation Act, included provisions that addressed historic preservation. The DOT Act is much more general than

Section 106 NHPA in that it refers to properties other than those listed in the Register.[23]

The more general language has allowed more properties and parklands to enjoy status as protected areas by this legislation, a policy developed early in its history. The United States Supreme Court ruled in the 1971 case *Citizens to Preserve Overton Park v. Volpe* that parklands could have the same protected status as "historic sites."[23]

Multiple Property Submission

A multiple property submission (MPS) is a thematic group listing of the National Register of Historic Places that consists of related properties that share a common theme and can be submitted as a group. Multiple property submissions must satisfy certain basic criteria for the group of properties to be included in the National Register.

The process begins with the Multiple Property Documentation Form, which acts as a cover document rather than the nomination to the National Register of Historic Places. The purpose of the documentation form is to establish the basis of eligibility for related properties. The information of the Multiple Property Documentation Form can be used to nominate and register related historic properties simultaneously, or to establish criteria for properties that may be nominated in the future. Thus, additions to an MPS can occur over time.

The nomination of individual properties in an MPS is accomplished in the same manner as other nominations. The name of the "thematic group" denotes the historical theme of the properties. It is considered the "multiple property listing." Once an individual property or a group of properties is nominated and listed in the National Register, the Multiple Property Documentation Form, combined with the individual National Register of Historic Places Nomination Forms, constitute a Multiple Property Submission.[26] Examples of MPS include the Lee County Multiple Property Submission, the Warehouses in Omaha, the Boundary Markers of the Original District of Columbia, and the Illinois Carnegie Libraries. Before the term "Multiple Property Submission" was introduced in 1984, such listings were known as "Thematic Resources" or "Multiple Resource Areas."[27]

Types of properties

Listed properties are generally in one of five broad categories, although there are special considerations for other types of properties that in any one, or into more specialized subcategories. The five general categories for National Register properties are: building, structure, site, district, and object.[20] In addition, historic districts consist of contributing and non-contributing properties.

Buildings, as defined by the National Register, are distinguished in the traditional sense. Examples include a house, barn, hotel, church, or similar construction. They are created primarily to shelter human activity. The term building, as in outbuilding, can be used to refer to historically and functionally related units, such as a courthouse and a jail or a barn and a house.[20]

Structures differ from buildings in that they are functional constructions meant to be used for purposes other than sheltering human activity. Examples include an aircraft, a grain elevator, a gazebo, and a bridge.

Objects are usually artistic in nature, or small in scale compared to structures and buildings. Although objects may be movable, they are generally associated with a specific setting or environment. Examples of objects include monuments, sculptures, and fountains.

Sites are the locations of significant events, which can be prehistoric or historic in nature and represent activities or buildings (standing, ruined, or vanished). When sites are listed, it is the locations themselves that are of historical interest. They possesses cultural or archaeological value regardless of the value of any structures that currently exist at the locations. Examples of types of sites include shipwrecks, battlefields, campsites, natural features, and rock shelters.[20]

Historic districts possess a concentration, association, or continuity of the other four types of properties. Objects, structures, buildings, and sites in a historic district are united historically or aesthetically, either by choice or by the nature of their development.[20]

There are several other different types of historic preservation associated with the properties of the National Register of Historic Places that cannot be classified as either simple buildings and historic districts. Through the National Park Service, the National Register of Historic Places publishes a series of bulletins designed to aid in evaluating and applying the criteria for evaluation of different types of properties.[20] Although the criteria are always the same, the manner they are applied may differ slightly, depending upon the type of property involved. The National Register bulletins describe application of the criteria for aids to navigation, historic battlefields, archaeological sites, aviation properties, cemeteries, and burial places, historic designed landscapes, mining sites, post offices, properties associated with significant persons, properties achieving significance within the last fifty years, rural historic landscapes, traditional cultural properties, and vessels and shipwrecks.[20]

Recent past

In American historic preservation, the **fifty-year rule** is the generally held belief that a property cannot be listed in the National Register of Historic Places unless it is at least fifty years old. Actually, this rule is not a hard rule at all; it is an exception to the National Register's criteria for evaluation. As stated by John H. Sprinkle, Jr., Deputy Director of the Federal Preservation Institute, "this 'rule' is only an exception to the criteria that shape listings within the National Register of Historic Places. Of the eight "exceptions" [or criteria considerations], Consideration G, for properties that have achieved significance within the past fifty years, is probably best-known, yet also misunderstood preservation principal in America."[28]

With each passing year a new group of resources crosses the fifty-year threshold, that were before considered "too recent" for listing on the National Register of Historic Places (unless they were exceptionally significant). The preservation of these "underage" resources, has gained increasing attention in recent years.

Limitations

The limitations of the NHPA are demonstrated when historic properties are destroyed, as when the Jobbers Canyon Historic District in downtown Omaha, Nebraska was demolished in 1987 to make way for a suburban-style corporate campus.[29] As of 1999, there have been 982 properties removed from the Register, most often due to being destroyed.[30]

Similar designations outside the United States

- Canada - National Historic Sites of Canada
- France - Monument historique
- Germany - Deutsche Stiftung Denkmalschutz and National Heritage Sites (Kulturdenkmal)
- Hong Kong - Historic building, see List of Grade I historic buildings in Hong Kong, List of Grade II historic buildings in Hong Kong and List of Grade III historic buildings in Hong Kong
- Italy - Monumento nazionale
- Netherlands - Rijksmonument
- Chile - National Monuments of Chile
- New Zealand - New Zealand Historic Places Trust
- Singapore - National Monuments of Singapore
- United Kingdom - Listed building or scheduled monument
- Portugal - IGESPAR
- Romania - National Register of Historic Monuments in Romania

See also

- African American Historic Places
- Architectural photographers
- Contributing property
- Cultural landscape
- Historic preservation
- History of the National Register of Historic Places
- Jails and prisons listed on the National Register of Historic Places
- Keeper of the Register
- List of heritage registers
- List of National Historic Landmarks by state
- List of threatened historic sites in the United States
- National Heritage Area
- National Historical Park
- Property type (National Register of Historic Places)
- Public history
- Register of Culturally Significant Property
- State Historic Preservation Office
- United States Memorials
- United States National Register of Historic Places listings
- World Heritage Site

References

Notes

[1] National Register Information System (http://www.nps.gov/nr/), National Register of Historic Places, *National Park Service*. Retrieved October 11, 2011.

[2] " National Park Service Directors and Directorate (http://www.cr.nps.gov/history/online_books/tolson/histlist6.htm)," Historic Listing of National Park Service Officials, *National Park Service*. Retrieved March 22, 2007

[3] National Historic Preservation Act of 1966 (http://www.cr.nps.gov/local-law/nhpa1966.htm), Public Law 102–575, *National Register of Historic Places*, Official site. Retrieved March 21, 2007.

[4] Mackintosh, Barry. " The Historic Sites Survey and National Historic Landmarks Program: A History (http://www.cr.nps.gov/history/online_books/mackintosh4/nhl.pdf)," (PDF), *National Historic Landmarks Program*, Official site. Retrieved March 23, 2007.

[5] Ferguson, T. J. " Native Americans and the Practice of Archaeology (http://links.jstor.org/sici?sici=0084-6570(1996)2:25<63:NAATPO>2.0.CO;2-B)," (JSTOR), *Annual Review of Anthropology*, Vol. 25. (1996), pp. 63–79. Retrieved March 23, 2007.

[6] Fisher, Charles E. " Promoting the Preservation of Historic Buildings: Historic Preservation Policy in the United States (http://links.jstor.org/sici?sici=0848-8525(1998)29:3/4<7:PTPOHB>2.0.CO;2-Z)," (JSTOR), *APT Bulletin*, Vol. 29, No. 3/4, Thirtieth-Anniversary Issue. (1998), pp. 7–11. Retrieved March 21, 2007.

[7] Scarpino, Philip V. " Planning for Preservation: A Look at the Federal-State Historic Preservation Program, 1966-1986 (in The Intergovernmental Politics of Preservation) (http://links.jstor.org/sici?sici=0272-3433(199221)14:2<49:PFPALA>2.0.CO;2-7)," (JSTOR) *The Public Historian*, Vol. 14, No. 2. (Spring, 1992), pp. 49-66. Retrieved March 21, 2007.

[8] Bearss, Edwin C. " The National Park Service and Its History Program: 1864-1986: An Overview (in The National Park Service and Historic Preservation) (http://links.jstor.org/sici?sici=0272-3433(198721)9:2<10:TNPSAI>2.0.CO;2-J)," (JSTOR), *The Public Historian*, Vol. 9, No. 2, The National Park Service and Historic Preservation. (Spring, 1987), pp. 10–18. Retrieved March 22, 2007.

[9] Hertfelder, Eric. " The National Park Service and Historic Preservation: Historic Preservation beyond Smokey the Bear (in Commentary: How Well Is the National Park Service Doing?) (http://links.jstor.org/sici?sici=0272-3433(198721)9:2<135:TNPSAH>2.0.CO;2-5)," (JSTOR), *The Public Historian*, Vol. 9, No. 2, The National Park Service and Historic Preservation. (Spring, 1987), pp. 135–142. Retrieved March 21, 2007.

[10] " About the National Register (http://www.cr.nps.gov/nr/about.htm)," National Register of Historic Places, *National Park Service*. Retrieved March 22, 2007.

[11] By contrast, the state of Colorado, for example, does not set any limits on owners of National Register properties. See "National and state registers," at Colorado Office of Archeology & Historic Preservation (http://coloradohistory-oahp.org/programareas/register/registers.htm)

[12] NPS.gov (http://www.nps.gov/hps/tps/tax/rhb/index.htm)

[13] " What are the results of a listing? (http://www.cr.nps.gov/nr/results.htm)," *National Register of Historic Places*, Official site. Retrieved March 21, 2007.

[14] " Historic Preservation Tax Incentives (http://www.cr.nps.gov/hps/tps/tax/index.htm)," Technical Preservation Services, *National Park Service*, Official site. Retrieved March 22, 2007.

[15] About the Historic Preservation Tax Incentives (http://www.cr.nps.gov/hps/tps/tax/brochure1.htm), Federal Historic Preservation Tax Incentives, Technical Preservation Services, *National Park Service*, Official site. Retrieved March 22, 2007.

[16] " Save America's Treasurers: FAQ (http://www.cr.nps.gov/hps/treasures/faqs.htm#2)" National Park Service. Retrieved March 23, 2007.

[17] " Department of Transportation Act (http://www.cr.nps.gov/local-law/FHPL_DOTAct.pdf)," (PDF), *National Park Service*, Official site. Retrieved March 21, 2007.

[18] " Listing a Property: Some Frequently Asked Questions (http://www.cr.nps.gov/nr/listing.htm)," *National Register of Historic Places*, Official site. Retrieved March 21, 2007.

[19] " Robie House (http://gis.hpa.state.il.us/hargis/PDFs/200493.pdf)," (PDF), National Register of Historic Places Nomination Form, *Illinois Historic Preservation Agency*. Retrieved April 11, 2007.

[20] " How to Apply the National Register Criteria for Evaluation (http://www.cr.nps.gov/nr/publications/bulletins/nrb15/nrb15.pdf)," (PDF), National Register Bulletins, *National Park Service*. Retrieved March 22, 2007.

[21] " Strengths of Local Listing (http://www.cr.nps.gov/hps/workingonthepast/strengths.htm)," Working on the Past in Local Historic Districts, *National Park Service*. Retrieved March 21, 2007.

[22] American Legation (http://tps.cr.nps.gov/nhl/detail.cfm?ResourceId=1839&ResourceType=Building), NHL Database, *National Historic Landmarks Program*. Retrieved April 11, 2007.

[23] Gray, Oscar S. " The Response of Federal Legislation to Historic Preservation (http://links.jstor.org/sici?sici=0023-9186(197122)36:3<314:TROFLT>2.0.CO;2-6)," (JSTOR), *Law and Contemporary Problems*, Vol. 36, No. 3, Historic Preservation. (Summer, 1971), pp. 314-328. Retrieved March 21, 2007.

[24] " Section 106 Summary (http://www.achp.gov/106summary.html)," *Advisory Council on Historic Preservation*, Official site, April 26, 2002.

[25] " Federal, State and Local Historic Districts (http://www.cr.nps.gov/hps/workingonthepast/toolbox1.htm), National Park Service. *Retrieved March 21, 2007.*

[26] Lee, Antoinette J. and McClelland Linda F. " How to Complete the National Register Multiple Property Documentation Form (http://www.cr.nps.gov/nr/publications/bulletins/nrb16b/nrb16b.pdf), (PDF), National Register Bulletin, *National Park Service*. Retrieved March 22, 2007.

[27] " Multiple Property Submission List (http://www.cr.nps.gov/nr/research/mpslist.htm)," National Register of Historic Places, *National Park Service*. Retrieved March 22, 2007.

[28] Sprinkle, John H., Jr (Spring 2007). ""Of Exceptional Importance": The Origins of the "Fifty-Year Rule" in Historic Preservation". *The Public Historian* **29** (2): 81–103. doi:10.1525/tph.2007.29.2.81.

[29] (1987) "Historic district at issue in Omaha," (http://query.nytimes.com/gst/fullpage.html?sec=travel&res=9B0DE1D81731F930A25751C1A961948260) *New York Times*. 12/13/87. Retrieved 7/8/07.

[30] Church sues over historic site (http://jacksonville.com/tu-online/stories/032399/met_2a1AME_1.html), Caren Burmeister, *Jacksonville Times-Union*, 3/23/99. Retrieved 7/8/11.

Bibliography

- Wiley, John. "National Register of Historic Places," *National Park Service*, (1994), ISBN 0-471-14403-7
- "National Register Information System" (http://www.nps.gov/nr/), *National Register of Historic Places*, National Park Service. Retrieved July 11, 2011.
- Shrimpton, Rebecca H., ed. " How to Apply the National Register Criteria for Evaluation (http://www.cr.nps.gov/nr/publications/bulletins/nrb15/)," *National Park Service*, (1997) National Register Bulletin No. 15, Washington, DC: Government Printing Office. Retrieved April 11, 2007.
- Title 36, U.S. Code of Federal Regulations (http://www.access.gpo.gov/cgi-bin/cfrassemble.cgi?title=200636), specifically Part 60 (http://www.access.gpo.gov/nara/cfr/waisidx_06/36cfr60_06.html), National Archives and Records Administration. Retrieved April 11, 2007.

External links

- National Register of Historic Places (http://www.nps.gov/nr/)
  - National Historic Landmarks Program (http://www.cr.nps.gov/nhl/INDEX.htm)
  - Travel itineraries (http://www.cr.nps.gov/nr/travel/), National Park Service
  - Teaching with Historic Place(TwHP) (http://www.nps.gov/history/NR/twhp/), National Park Service
  - Weekly updates (http://www.cr.nps.gov/nr/nrlist.htm)
  - Nomination forms (http://www.cr.nps.gov/nr/publications/forms.htm)
- Advisory Council on Historic Preservation (http://www.achp.gov/)
  - Working with Section 106 (http://www.achp.gov/work106.html), Advisory Council on Historic Preservation
  - An Overview of Federal Historic Preservation Law, 1966-1996 (http://www.achp.gov/book/TOC2.html#Table of Contents)

Round_barn

A **round barn** is a historic barn design that could be octagonal, polygonal, or circular in plan. Though round barns were not as popular as some other barn designs, their unique shape makes them noticeable. The years from 1880–1920 represent the height of round barn construction.[1] Round barn construction in the United States can be divided into two overlapping eras. The first, the octagonal era, spanned from 1850–1900. The second, the true circular era, spanned from 1889–1936. The overlap meant that round barns of both types, polygonal and circular, were built during the latter part of the nineteenth century.[2] Numerous round barns in the United States are listed on the National Register of Historic Places.[3]

The round barn at Hancock Shaker Village

History

Round barns date to the 18th and early 19th century. George Washington designed and built a sixteen-sided threshing barn at his Dogue Run Farm in Fairfax County, Virginia in 1793.[4] It is considered the first American round barn. Early round barns were particularly associated with the Shaker community,[1] [5] one was constructed in 1826 at the Shaker community in Hancock, Massachusetts.[1] A few other round barns appeared on the American landscape before the Civil War.[4]

Despite considerable publicity of the 1826 Shaker barn, the design did not become popular until the 1880s, when some agricultural colleges began to push the design as they taught progressive farming methods, based on the principles of industrial efficiency. It was from 1880–1920 that round barns began were the most popular in the United States, especially in the Midwest.[1]

The rise in popularity and the promotion of round barns occurred surrounding the new focus on efficiency. The circular shape has a greater volume-to-surface ratio than a square barn. Regardless of size, this made round barns cheaper to construct than similar-sized square or rectangular barns because they required less materials. The structural stability is also enhanced over that of a typical quadrilaterally shaped barn. Simplified construction lacking elaborate truss systems for the arched roof was also seen as an advantage. In the Midwest, particularly, the buildings were thought more resilient against prairie thunderstorms.[5] The interior layout of round barns was promoted as more efficient, since farmers could work in a continuous direction.[1] In the days before mechanization, labor-saving features were a big selling point.

The interest in round barns spread to California in the later 19th century and several were built there. Santa Rosa, California is home to two well-preserved and well-known round barns.[6] One, the Fountaingrove Round Barn, is located on land that was part of the Fountain Grove spiritual commune, the Brotherhood of New Life, founded by Thomas Lake Harris around 1875.[6] Built near the end of the 19th century, it was part of the Fountaingrove winery owned by Harris's protégé, Kanawe Nagasawa, who reportedly designed the structure. Another survivor is the De Turk Round Barn on Donahue Street, and was built in the late 1870s by local settler and businessman Isaac De Turk.[7]

Fountaingrove Round Barn in Santa Rosa

Claims of round barn efficiency were overstated. The round barn never caught on as a standard barn, as some of those pushing the progressive, efficiency-based agricultural methods had hoped. The spread of machinery, especially with the Rural Electrification program, eliminated the advantages of labor-saving designs that were more complicated to build, and the popularity of round barns faded. Regardless, numerous round barns were constructed during the period of popularity the design enjoyed, and a large number still stand today.[1]

Round barn era

Octagonal

The "Octagonal Era" of round barn design stretched from about 1850 until 1900. Round barns, such as Washington's, were often multi sided in their earliest incarnations. Multi-sided round barns came in a variety of polygonal shapes, including layouts of six, eight, nine, ten, twelve, fourteen and sixteen sides. Polygonal barns constructed before the advent of balloon framing tended to have interior spaces that were more rectangular than circular.[2]

Tim Thering Octagon Barn (Plain Wisconsin)

True circular

The "True Circular Era" of round barn construction spanned from 1889–1936, overlapping the octagonal era and finally dwindling out as round barns fell out of popularity. True circular round barns began to rise as improvements in construction techniques made their design more practical. As balloon framing, circular silos and truly self-supporting roofs were developed, circular barns superseded polygonal structures and began to be built in greater number. Despite the gains in popularity for circular barns, polygonal barns continued to be built up through the height of the True Circular Era.[2]

A true circular barn in Minnesota

End of an era

By the 1920s round barn construction had begun to decline in some locations. In Illinois, the popularity of round barns was in part due to the University of Illinois round barns, and round barn construction had tapered off considerably. Several reasons have been given for the decline in the popularity of round barn designs. The standardization of the construction industry and the resulting decline in timber framing following the American Civil War is one possible reason. Another possibility is that the mechanization of American agriculture was more suited to rectangular barn design.[8]

Design

Designed in a distinctive circular shape, many these barns were meant to take advantage of gravity to move hay from the loft to the cow stable below. In many cases, a silo was constructed to rise up through the round barn's center. The round barn was promoted as a labor-saving design by agricultural colleges as a progressive way to house dairy cattle.[9]

Interior view of the Pete French Round Barn near Frenchglen, Oregon

In the case of the Pete French Round Barn near Frenchglen, Oregon, the barn was built with a rock wall around an inner stable area, and included a covered run around the wall where horses could be exercised during the harsh high desert winters.

The earliest round barns tend to have several flat sides, usually twelve or sixteen. They also tend to be wood-sided while the later round barns are more often faced with brick or glazed tile. The interior design of round barns shifted as well. The early round barns had cattle stanchions on the first floor with the whole of the loft used for hay and feed storage. Later barns possessed a central space which rose up from the ground level through the entire building. The cattle stanchions in this variation of round barn were arranged around a circular manger on the lower level. Above the stanchion level a circular wagon drive allowed hay to be unloaded into the central mow as the wagon circled the perimeter. The final stage of interior design in round barns included a storage silo through the center of the structures. These were not really added until silos became fixtures of American farms. Sometimes the central silo would project up through the roof.[1]

Influences

It is known that prominent agricultural colleges began to promote the design technique as round barns came to prominence. However, the provider of the initial impetus is the subject of some debate. In 1848 Orson Fowler published *A Home For All: Of the Gravel Wall and Octagon Mode of Building,* extolling the virtues of the octagonal shape in home construction. The book included a discussion on use of the shape in barns and other outbuildings. The book generated a flurry of octagonal home construction, especially in the New England and Middle Atlantic states. Some researchers have linked the earliest round barn constructions with Fowler's popular book because so many of the early round barns assumed the octagonal shape.[4]

Ten-sided barn near Mauston, Wisconsin

Other historians discount Fowler's influence on the beginning of the round barn era in the United States. Soike asserts that Fowler did not have any direct connection with any octagonal round barns. He noted that the octagon had ceased to be the basis for building constructions by the Civil War.[4] In Indiana, for example, 219 round barns were constructed between 1850 and 1936; of those, 67 were polygonal, including 17 eight-sided barns built after 1900.[2]

In the Midwest, especially in Illinois, the round barns at the University of Illinois led to an increase in the style's popularity statewide.[8] This was partly due to the U of I's Agricultural Experiment Station's publishing regular "Bulletins". The *Agricultural Experiment Station Bulletin* coupled with an article by H.C. Crouch touting round barns in the *Illinois Agriculturalist* led to the construction of round barns across the state. Anecdotal evidence of the impact of the University of Illinois round barns can be collected from farmers today. Stories about fathers and grandfathers recollect round barns being constructed on account of what was going on "over at the University".[8]

See also

- List of round barns
- Barn
- Lewis Round Barn
- Neff Round Barn
- Round Barns in Illinois Thematic Resources
- Thomas Ranck Round Barn
- San Luis Obispo Octagon Barn

Notes

[1] Auer, Michael J. The Presevation of Historic Barns (http://www.cr.nps.gov/hps/tps/briefs/brief20.htm), Preservation Briefs, National Park Service (October 1989). Retrieved 7 February 2007

[2] Round and Polygonal Barns of Indiana (http://pdfhost.focus.nps.gov/docs/NRHP/Text/64500218.pdf), (PDF), National Register of Historic Places Multiple Property Documentation Form, NRIS Database, National Register of Historic Places. Retrieved 8 February 2007

[3] "National Register Information System" (http://nrhp.focus.nps.gov/natreg/docs/All_Data.html). *National Register of Historic Places*. National Park Service. 2009-03-13. .

[4] North Dakota Round Barns (http://pdfhost.focus.nps.gov/docs/NRHP/Text/64000477.pdf) (PDF), National Register of Historic Places Inventory Nomination Form, NRIS Database, National Register of Historic Places. Retrieved 8 February 2007

[5] Round Barns (http://uitours.ncsa.uiuc.edu/landmarks/roundbarns/), University of Illinois Campus Tours, University of Illinois. Retrieved 7 February 2007

[6] " Fountain Grove (http://ci.santa-rosa.ca.us/default.aspx?PageId=69)," *City of Santa Rosa*, 2004-2006, official site. Retrieved 16 September 2007

[7] "National Register Information System" (http://nrhp.focus.nps.gov/natreg/docs/All_Data.html). *National Register of Historic Places*. National Park Service. 2010-07-09. .

[8] University of Illinois Experimental Dairy Farm Historic District (http://gis.hpa.state.il.us/hargis/PDFs/201105.pdf), (PDF), National Register of Historic Places Registration Form, HAARGIS Database, Illinois Historic Preservation Agency. Retrieved 19 April 2007

[9] Historic Barn Types (http://www.uvm.edu/~vhnet/hpres/publ/barnb/bbhbty.html), Taking Care of Your Old Barn, University of Vermont, Vermont Division for Historic Preservation. Retrieved 7 February 2007

External links

- Efficiency of the Round barn (http://www.thebarnjournal.org/round/efficiency.html): Kansas State Board of Agriculture, Eighteenth Biennial Report, 1911–12, pp. 139–42.
- Shaker Round Barn (http://www.hancockshakervillage.org/experience/stone_barn.html): Official site
- Sleuthing for Clues (http://www.mountvernon.org/visit/plan/index.cfm/pid/593/): article covering reconstruction effort of Washington's round barn. Mount Vernon Estate and Gardens (http://www.mountvernon.org/)

Burns,_Oregon

Burns, Oregon

— City —

North Broadway Avenue

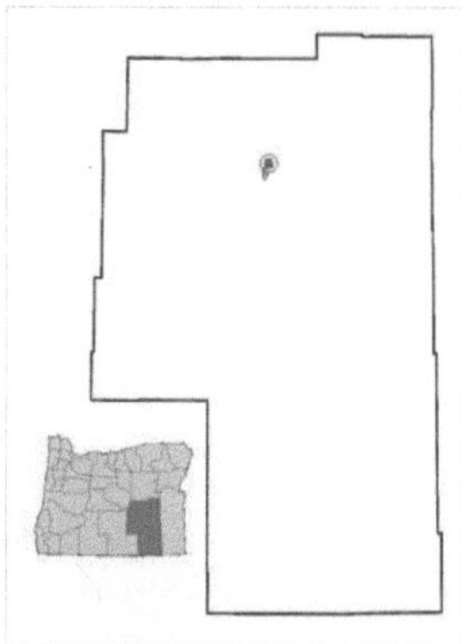

Location in Oregon

Coordinates: 43°35′13″N 119°3′36″W

Country	United States
State	Oregon
County	Harney
Incorporated	1889
Government	
• **Mayor**	Craig LaFollette[1]
Area[2]	
• **Total**	3.55 sq mi (9.19 km^2)
• **Land**	3.55 sq mi (9.19 km^2)
• **Water**	0 sq mi (0 km^2)
Elevation	4147 ft (1264 m)
Population (2010)[3]	
• **Total**	2806
• **Estimate** (2011[4])	2789
• **Density**	790.4/sq mi (305.2/km^2)
	U.S. Census
Time zone	Pacific (UTC-8)
• **Summer (DST)**	Pacific (UTC-7)

ZIP Code	97720
Area code(s)	541
FIPS code	41-09800[5]
GNIS feature ID	1118337[6]
Website	www.ci.burns.or.us [7]

Burns is a city in and the county seat of Harney County, Oregon, United States.[8] As of the 2010 census the population was 2,806.[9]

History

Burns was established in the early 1880s and incorporated upon Harney county's creation in 1889. It was named for the Scottish poet Robert Burns by early settler and County Commissioner George McGowan.[10]

Geography

Burns is located near the northern edge of the arid Harney Basin.

According to the United States Census Bureau, the city has a total area of 3.55 square miles (9.19 km^2), all of it land.[2] It is 136 miles (219 km) east of Bend, Oregon, and 130 miles (209 km) west of Ontario, Oregon, on U.S. Route 20 at its intersection with U.S. Route 395, 186 miles (299 km) south of Pendleton, Oregon. A third highway, Oregon Route 78, runs 92 miles (148 km) from Burns southeast past the Malheur National Wildlife Refuge and Steens Mountain to Burns Junction, Oregon.

Climate

Burns has a high-desert climate that averages 115 clear days a year and only 29 rainy days. Temperatures average 25 °F (–4 °C) in January and 85 °F (29 °C) in July. Annual precipitation is about 11 inches (28 cm), and about 48 inches (122 cm) of snow a year falls in Burns.[11]

Monthly Normal and Record High and Low Temperatures for Burns, Oregon

Month	Jan	Feb	Mar	Apr	May	Jun	Jul	Aug	Sep	Oct	Nov	Dec
Rec High °F	57	67	74	84	94	98	107	100	97	86	70	57
Avg High °F	34.7	40.5	49	57.4	66.1	75.1	85.4	84.5	75	62.4	44.8	35.1
Avg Low °F	14	19.4	24.9	28.6	35.6	41.1	46.4	43.9	35	26.4	20.6	14.6
Rec Low °F	-27	-28	-14	10	15	21	25	22	17	-7	-13	-20
Precip (in)	1.18	1.11	1.24	0.85	1.05	0.66	0.4	0.45	0.5	0.72	1.11	1.3

Source: USTravelWeather.com[12]

Demographics

Broadway Street in Burns, Oregon.

Historical populations		
Census	Pop.	%±
1890	264	—
1900	547	107.2%
1910	904	65.3%
1920	1022	13.1%
1930	2599	154.3%
1940	2566	−1.3%
1950	3093	20.5%
1960	3523	13.9%
1970	3293	−6.5%
1980	3579	8.7%
1990	2913	−18.6%
2000	3064	5.2%
2010	2806	−8.4%
source:[13] [14]		

2010 census

As of the census[3] of 2010, there were 2,806 people, 1,280 households, and 720 families residing in the city. The population density was 790.4 inhabitants per square mile (305.2 /km^2). There were 1,490 housing units at an average density of 419.7 per square mile (162.0 /km^2). The racial makeup of the city was 92.2% White, 0.3% African American, 2.6% Native American, 0.7% Asian, 0.7% from other races, and 3.3% from two or more races. Hispanic or Latino of any race were 4.7% of the population.

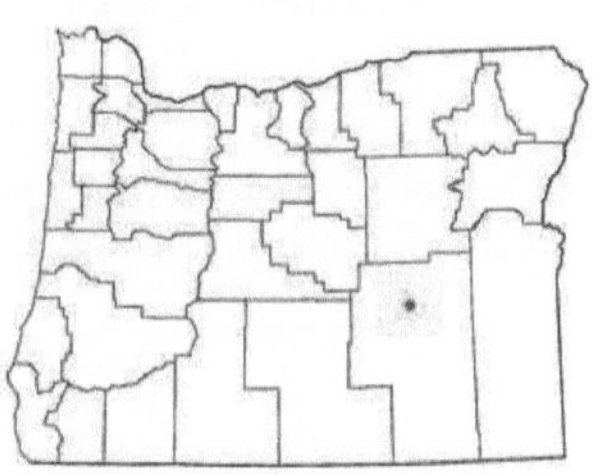

Burns' location in Oregon (red dot).

There were 1,280 households out of which 24.4% had children under the age of 18 living with them, 40.6% were married couples living together, 11.5% had a female householder with no husband present, 4.1% had a male householder with no wife present, and 43.8% were non-families. 36.9% of all households were made up of individuals and 13.9% had someone living alone who was 65 years of age or older. The average household size was 2.13 and the average family size was 2.77.

The median age in the city was 44.5 years. 21.5% of residents were under the age of 18; 8.7% were between the ages of 18 and 24; 20.3% were from 25 to 44; 30.7% were from 45 to 64; and 18.9% were 65 years of age or older. The gender makeup of the city was 51.2% male and 48.8% female.

2000 census

As of the census[5] of 2000, there were 3,064 people, 1,272 households, and 783 families residing in the city. The population density was 861.5 people per square mile (332.3/km²). There were 1,487 housing units at an average density of 418.1 per square mile (161.3/km²). The racial makeup of the city was 93.54% White, 0.20% African American, 2.48% Native American, 0.52% Asian, 0.10% Pacific Islander, 1.11% from other races, and 2.06% from two or more races. Hispanic or Latino of any race were 5.29% of the population. There were 1,272 households out of which 28.4% had children under the age of 18 living with them, 48.8% were married couples living together, 8.2% had a female householder with no husband present, and 38.4% were non-families. 31.9% of all households were made up of individuals and 12.4% had someone living alone who was 65 years of age or older. The average household size was 2.33 and the average family size was 2.94.

In the city the population dispersal was 26.0% under the age of 18, 7.4% from 18 to 24, 26.6% from 25 to 44, 23.7% from 45 to 64, and 16.3% who were 65 years of age or older. The median age was 39 years. For every 100 females there were 100.4 males. For every 100 females age 18 and over, there were 96.1 males.

The median income for a household in the city was $26,658, and the median income for a family was $31,792. Males had a median income of $24,858 compared to $22,097 for females. The per capita income for the city was $16,224. About 8.4% of families and 12.3% of the population were below the poverty line, including 8.7% of those under age 18 and 15.1% of those age 65 or over.

Politics

The Burns Paiute Tribe Indian Colony of Oregon is a federally recognized tribe of Northern Paiute Native Americans that operates their tribal headquarters in Burns.[15]

Transportation

Air

Burns Municipal Airport provides general aviation services but currently has no scheduled flights.[16]

Bus

- The Eastern Point is an intercity bus offering service between Bend and Ontario, Oregon that stops in Burns and provides connections in Bend to the Central Oregon Breeze and Porter Stage Lines.[17]

Notable people

- Kellen Clemens - NFL quarterback - St. Louis Rams
- Norma Paulus - Former Oregon politician
- Robert Freeman Smith - Congressman/Speaker of Oregon House

Media

The weekly *Burns Times-Herald* is the only newspaper in the city.[18] Two radio stations, KORC (92.7 FM) and KBNH (1230 AM), broadcast from Burns.[19]

Harney Basin, near Burns, Oregon.

References

[1] http://burnstimesherald.info/2011/04/20/burns-appoints-craig-lafollette-as-mayor-fills-two-council-seats/

[2] "US Gazetteer files 2010" (http://www.census.gov/geo/www/gazetteer/files/Gaz_places_national.txt). United States Census Bureau (http://en.wikipedia.org/wiki/United_States_Census_Bureau). . Retrieved 2012-12-21.

[3] "American FactFinder" (http://factfinder2.census.gov/faces/nav/jsf/pages/index.xhtml). United States Census Bureau (http://en.wikipedia.org/wiki/United_States_Census_Bureau). . Retrieved 2012-12-21.

[4] "Population Estimates" (http://www.census.gov/popest/data/cities/totals/2011/files/SUB-EST2011-IP.csv). United States Census Bureau (http://en.wikipedia.org/wiki/United_States_Census_Bureau). . Retrieved 2013-01-04.

[5] "American FactFinder" (http://factfinder.census.gov). United States Census Bureau. . Retrieved 2008-01-31.

[6] "US Board on Geographic Names" (http://geonames.usgs.gov). United States Geological Survey. 2007-10-25. . Retrieved 2008-01-31.

[7] http://www.ci.burns.or.us/

[8] "Find a County" (http://www.naco.org/Counties/Pages/FindACounty.aspx). National Association of Counties. . Retrieved 2011-06-07.

[9] http://factfinder2.census.gov/faces/tableservices/jsf/pages/productview.xhtml?pid=DEC_10_PL_GCTPL2.ST10&prodType=table

[10] "Burns, Oregon" (http://web.archive.org/web/20070928072003/http://www.harneycounty.com/DowntownBurns/Burns.htm). Harney County Chamber of Commerce. Archived from the original (http://www.harneycounty.com/DowntownBurns/Burns.htm) on 2007-09-28. . Retrieved 2007-10-30.

[11] "Weatherbase: Historical Weather for Burns, Oregon, United States of America" (http://www.weatherbase.com/weather/weather.php3?s=038627&refer=). Weatherbase. 2007. . Retrieved 2007-11-01.

[12] "Historical Burns Weather Data" (http://web.archive.org/web/20070929103223/http://www.ustravelweather.com/weather-oregon/burns-weather.asp). USTravelWeather.com. Archived from the original (http://www.ustravelweather.com/weather-oregon/burns-weather.asp) on 2007-09-29. . Retrieved 2007-11-01.

[13] Moffatt, Riley. *Population History of Western U.S. Cities & Towns, 1850-1990.* Lanham: Scarecrow, 1996, 207.

[14] "Subcounty population estimates: Oregon 2000-2007" (http://www.census.gov/popest/cities/files/SUB-EST2007-41.csv) (CSV). United States Census Bureau, Population Division. 2009-03-18. . Retrieved 2009-05-09.

[15] "The Old Camp Casino." (http://500nations.com/casinos/orOldCamp.asp) *500 Nations.* 2009 (retrieved 8 Dec 2009)

[16] Burns Airport FAA Form 5010 (http://gcr1.com/5010web/airport.cfm?Site=BNO)

[17] "Eastern Point Schedule" (http://www.highdesert-point.com/assets/Schedule-Docs/eastern-pointSchedule11.pdf). TAC Transportation. . Retrieved July 24, 2011.

[18] Friedman, Ralph (July 10, 1960). "Old West lingers on in Oregon's High Country" (http://select.nytimes.com/gst/abstract.html?res=F00E13FC385A1A7A93C2A8178CD85F448685F9). *The New York Times*: p. XX22. . "The county has only one movie house, in Burns; only two physicians, both in Burns; only two television sets outside of Burns and one weekly newspaper, The "Burns Times Herald"."

[19] "Oregon Blue Book: Media" (http://bluebook.state.or.us/cultural/media/media03.htm). Oregon State Archives. . Retrieved 2007-11-01.

External links

- Entry for Burns (http://bluebook.state.or.us/local/cities/ad/burns.htm) in the *Oregon Blue Book*
- Profile of Burns from the Harney County Chamber of Commerce (http://www.harneycounty.com/DowntownBurns/Burns.htm)

Harney_County,_Oregon

Harney County, Oregon	
 Harney County Courthouse in Burns	
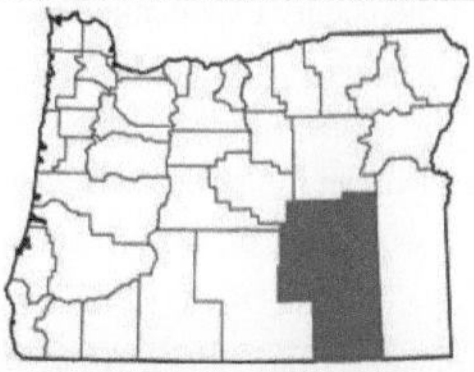 Location in the state of Oregon	
Oregon's location in the U.S.	
Founded	1889
Seat	Burns
Area **- Total** **- Land** **- Water**	10226 sq mi (26485 km²) 10134 sq mi (26247 km²) 92 sq mi (238 km²), 0.90%
Population **- (2010)** **- Density**	7422 1/sq mi (.28/km²)
Website	www.co.harney.or.us [1]

Harney County is a county located in the U.S. state of Oregon. It is included in the eight-county definition of Eastern Oregon. Established in 1889, the county is named in honor of William S. Harney, a military officer of the period, who was involved in the Pig War and popular in the Pacific Northwest. The county seat is Burns. In 2010, the county's population was 7,422. Harney County is Oregon's largest county in area.

History

The Native Americans living in this region at the time of the Lewis and Clark Expedition were the Northern Paiute, who fought with the Tenino and Wasco peoples. Peter Skene Ogden was the first known European to explore this area in 1826 when he led a fur brigade for the Hudson's Bay Company.

Harney County was carved out of the southern two-thirds of Grant County on February 25, 1889. A fierce political battle, with armed "night riders" who spirited county records from Harney to Burns, ended with Burns as the county seat in 1890.

The Malheur River Indian Reservation was created by executive order on March 14, 1871 and the Northern Paiute within the Oregon state boundaries were settled there. Descendants of these people form a federally recognized tribal entity, the Burns Paiute Tribe, which has about 341 members.[2] Fewer than 35.5% of the tribal members live on the Burns Paiute Indian Colony near Burns.[2] The tribe earns revenue from a small casino, the Old Camp Casino, and renting out communal tribal lands for grazing rights to local ranchers.

Economy

Three industries have traditionally provided the county's economic base: ranching, sheep raising, and timber. The railroad, which extended into the area in 1883, served as a catalyst to the cattle industry but later contributed to its decline. By bringing farmers and sheep men to the area, it created increased competition for productive land. Harvesting and breeding of wild horses was lucrative for a period. Harney County shares the largest Ponderosa Pine forest in the nation with Grant County. Its abundance of game, numerous campsites and excellent fishing have stimulated fast-growing recreational activities.

Although county lands were open to homesteading from 1862 to 1934, the U.S. Bureau of Land Management still owns more than three million acres (12,000 km^2), or 62% of the lands within the county boundaries. Facilitated on the national level by the Carey act of 1894, arid land in Harney County was donated to the state for irrigation and settlement, but all water development efforts failed.

Eventually all land claims filed under the reclamation legislation were abandoned or nullified. Malheur National Wildlife Refuge was established in 1908 and expanded in 1936. The refuge now includes **unknown operator: u','** acres (**unknown operator: u'strong'** km^2). Borax has been mined in the Steens area, and uranium has been found on its south side.

Politics

Like all counties in eastern Oregon, the majority of registered voters who are part of a political party in Harney County are members of the Republican Party. In the 2008 presidential election 70.45% of Harney County voters voted for Republican John McCain, while 25.79% voted for Democrat Barack Obama and 3.73% of voters either voted for a Third Party candidate or wrote in a candidate.[3] These numbers show a slight shift towards the Democratic candidate when compared to the 2004 presidential election, in which 76% of Harney Country voters voted for George W. Bush, while 22.7% voted for John Kerry, and 1.3% of voters either voted for a Third Party candidate or wrote in a candidate.[4]

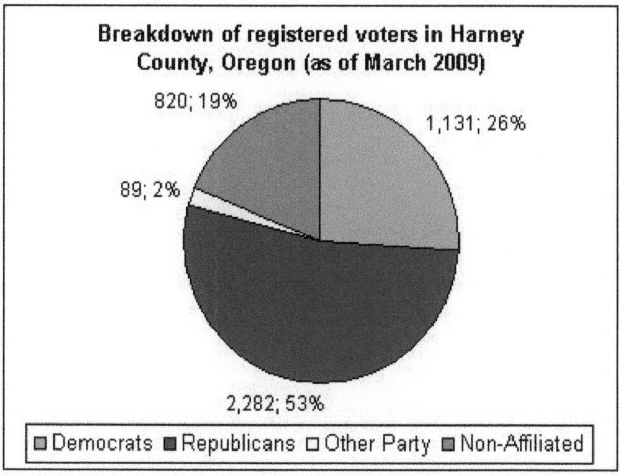

[5]

Geography

Sign welcoming drivers to Harney County

According to the U.S. Census Bureau, the county has a total area of 10226 square miles (26485.2 km^2), of which 10134 square miles (26246.9 km^2) is land and 92 square miles (238.3 km^2) is water, mostly as part of Malheur Lake. The total area is 0.90% water.

Harney is the largest county in Oregon and the ninth largest county in the United States (excluding boroughs and census areas in Alaska).

Steens Mountain is the county's most prominent geographical feature, rising 9700 feet (m) above sea level and spanning many miles across a region that is otherwise fairly flat. To its southeast is the Alvord Desert, the driest place in Oregon, and to the north lies Harney Basin, which contains Malheur Lake and Harney Lake.

Adjacent counties

- Lake County, Oregon - (west)
- Deschutes County, Oregon - (northwest)
- Crook County, Oregon - (northwest)
- Grant County, Oregon - (north)
- Malheur County, Oregon - (east)
- Humboldt County, Nevada - (south)
- Washoe County, Nevada - (southwest)

National protected areas

- Malheur National Forest (part)
- Malheur National Wildlife Refuge
- Ochoco National Forest (part)

Demographics

Historical populations		
Census	Pop.	%±
1890	2559	—
1900	2598	1.5%
1910	4059	56.2%
1920	3992	−1.7%
1930	5920	48.3%
1940	5374	−9.2%
1950	6113	13.8%
1960	6744	10.3%
1970	7215	7.0%
1980	8314	15.2%
1990	7060	−15.1%
2000	7609	7.8%
2010	7422	−2.5%
[6] [7] [8]		

As of the census[9] of 2000, there were 7,609 people, 3,036 households, and 2,094 families residing in the county. The population density was 1 people per square mile (0/km²). There were 3,533 housing units at an average density of 0 per square mile (0/km²). The racial makeup of the county was 91.93% White, 3.97% Native American, 0.51% Asian, 0.13% Black or African American, 0.07% Pacific Islander, 1.30% from other races, and 2.09% from two or more races. 4.15% of the population were Hispanic or Latino of any race. 21.1% were of German, 11.1% American, 10.3% Irish and 9.7% English ancestry according to Census 2000.

There is a small, but significant Spanish Basque community.[10]

Approximately 75% of the population of Harney County lives in the Burns-Hines municipal district. Crane is the only other localised population center, with less than 7% of the population of Harney County. Lawen and Riley have no localised populations. The remaining population of Harney County is dispersed throughout the countryside, mostly dwelling on large ranches.

There were 3,036 households out of which 29.40% had children under the age of 18 living with them, 58.00% were married couples living together, 6.80% had a female householder with no husband present, and 31.00% were non-families. 25.90% of all households were made up of individuals and 10.20% had someone living alone who was 65 years of age or older. The average household size was 2.45 and the average family size was 2.94.

In the county, the population was spread out with 26.00% under the age of 18, 6.40% from 18 to 24, 26.60% from 25 to 44, 26.10% from 45 to 64, and 15.00% who were 65 years of age or older. The median age was 40 years. For every 100 females there were 102.90 males. For every 100 females age 18 and over, there were 98.20 males.

The median income for a household in the county was $30,957, and the median income for a family was $36,917. Males had a median income of $27,386 versus $21,773 for females. The per capita income for the county was $16,159. About 8.60% of families and 11.80% of the population were below the poverty line, including 12.70% of those under age 18 and 13.90% of those age 65 or over.

Communities

Incorporated cities

- Burns
- Hines

Unincorporated communities

- Andrews
- Buchanan
- Crane
- Denio
- Diamond
- Drewsey
- Dunnean
- Fields
- Frenchglen
- Frost Mill
- Harney
- Indian Village
- Lawen
- Narrows
- New Princeton
- Riley
- Suntex
- Trout Creek
- Van
- Venator
- Voltage
- Wagontire
- Whitehorse

See also

- National Register of Historic Places listings in Harney County, Oregon

References

[1] http://www.co.harney.or.us

[2] Wadatika Today (http://www.burnspaiute-nsn.gov/index.php?option=com_content&view=article&id=64:wadatika-today&catid=37:history&Itemid=57)

[3] http://co.harney.or.us/PDF_Files/County%20Clerk/Election%20Results%20Update%202008/Eection%20Results%20Nov%204%202008%20General.pdf Retrieved on 4/21/09

[4] http://www.city-data.com/county/Harney_County-OR.html Retrieved on 4/21/09

[5] http://www.sos.state.or.us/elections/votreg/mar09.pdf Retrieved 4/21/09

[6] http://www.census.gov/population/www/censusdata/cencounts/files/or190090.txt

[7] http://factfinder2.census.gov

[8] http://mapserver.lib.virginia.edu/

[9] "American FactFinder" (http://factfinder.census.gov). United States Census Bureau. . Retrieved 2008-01-31.

[10] Etulain, Richard W.. "Basques" (http://oregonencyclopedia.org/entry/view/basques/). *The Oregon Encyclopedia*. . Retrieved 2008-04-03.

Lake_County,_Oregon

Lake County, Oregon	
Location in the state of Oregon	
Oregon's location in the U.S.	
Founded	October 24, 1874
Seat	Lakeview
Area **- Total** **- Land** **- Water**	8358 sq mi (21647 km²) 8136 sq mi (21072 km²) 223 sq mi (578 km²), 2.66%
Population **- (2010)** **- Density**	7895 1/sq mi (.37/km²)
Website	www.lakecountyor.org [1]

Lake County is a county in the high desert south-central region of the U.S. state of Oregon, so named for the many lakes found within its boundaries,[2] including Lake Abert, Hart Lake Reservoir, and Goose Lake. While Lake is among Oregon's largest counties by area, it is sparsely populated, with 7,895 residents as of 2010. Its seat is at Lakeview.[3]

Lake County is in a high desert region known as the Oregon Outback, on the northwestern edge of the Great Basin. The county is generally divided between the communities around Lakeview and Paisley to the south and the communities around Christmas Valley, Fort Rock, and Silver Lake to the north.

Its economy consists largely of agriculture and natural resource management and extraction. It is home to many large cattle ranches, hay farms, and timber holdings (both public and private), as well as several frontier towns and early 20th-century homesteads. Although lumber was once a primary economic driver in Lake County, today only one mill remains, at Lakeview.

History

Pre-Columbian

Pre-Clovis era coprolites found in the Paisley Caves in northern Lake County in 2007 have been radiocarbon dated to 14,300 calendar years before present. DNA extracted from these human remains bears certain genetic markers found only in Native American populations.[4] Luther Cressman found prehistoric artifacts in the Fort Rock Caves of northern Lake County in 1938, including basketry, stone tools,[5] and a cache of woven sagebrush bark sandals which have been dated to more than 10,000 years ago.[6]

Modern

European traders, explorers and military expeditions arrived in the region during the early part of the 19th century. Peter Skene Ogden led Hudson's Bay Company trappers to Goose Lake in 1827. In 1832, the Hudson Bay trappers under John Work were in the Goose Lake Valley and their journals mentioned Hunter's Hot Springs. Work's expedition visited Warner Lakes and Lake Abert and camped at Crooked Creek in the Chandler Park area. There they documented eating wild plums, which still grow in the area. They also reported being attacked by Indians.[7] In

1838, Colonel J. J. Abert, a U.S. engineer, prepared a map that includes Warner Lakes and other natural features using information from the Hudson Bay trappers. In 1843, John C. Fremont led a party which named Christmas (Hart) Lake.[7]

Lake County once hosted significant populations of Basque[8] and Irish sheepherders.[9] Disputes over grazing rights, exacerbated by the introduction of wheat farming, led to the eruption of range wars between cattle ranchers and sheep herders. At least one band of masked rifle-armed cattlemen killed sheep in the northern part of the county and in Deschutes County during the early 20th century and they came to be known as "sheepshooters". According to the Oregon History Project, 2,300 sheep were killed in a single night in April 1904 in Lake County.[10]

Lake County grew with the arrival of homesteaders, but the dry climate made for challenging development.

Lake County was created from Jackson and Wasco Counties on October 24, 1874, by the State Legislature. It then included the present Klamath County and all of the present Lake County except Warner Valley. In 1882, land was assigned to create Klamath County, and in 1885 the Warner area from Grant County was added. Linkville, now Klamath Falls, was the first county seat.[11]

M. Bullard gave 20 acres (80,000 m^2) as the Lakeview townsite. By the 1875 election, a town had been started and an election moved the county seat to Lakeview. Because of poor transportation connections with the rest of Oregon, the early economic orientation of Lake County was toward California: both the *San Francisco Chronicle* and the *San Francisco Examiner* arrived in Lakeview daily, often before *The Oregonian*. During the 1840s and 1850s the county was part of the military courier route between The Dalles on the Columbia River and the Presidio in San Francisco.

The county acquired a railroad connection in the 1890s. That railroad spur, the Nevada-California-Oregon Railway line running from Lakeview to Reno, Nevada, emphasized the isolation of the county from the rest of Oregon.[12] A devastating fire in 1900 destroyed much of Lakeview, including 75 businesses.[7]

Economy

The Lake County Courthouse

The economy in Lake County is reliant on lumber, agriculture, natural resource extraction, health care, a prison[13] and government. The area is also trying to promote itself for the many outdoor recreational and sightseeing opportunities offered.[14]

Lake County includes numerous cattle ranches. Irrigation has also permitted some agriculture based upon the raising of livestock and the growing of hay and grain despite the low rainfall and a short growing season.

Lumber and wood products are taken from the Fremont National Forest, Bureau of Land Management properties and private landholdings. The Collins Companies operates the last remaining mill in the area, the Lakeview sawmill, and is also a large landowner in the region.[15] Over 78% of the land in Lake County is owned and managed by the federal and state government.[16]

The Lakeview area of Lake County also includes a perlite mine[17] and once included uranium mining[18] (now subject to clean-up operations).[19] Several exploratory wells were dug for oil, but without success.

A railroad line ships timber products and perlite to Burlington Northern's rail hub in Alturas, California. At one time the railway was the only County owned and operated rail line in the country.

Government employees from the national forest and the regional Bureau of Land Management headquarters provide many of the higher paying salary job in an economy that otherwise would have to rely only on seasonal agricultural, tourism and lumber jobs. The Bureau of Land Management is landowner of 49% of the lands within the county.

Tourism is a growing industry because of the county's many attractions which include Hart Mountain National Antelope Refuge, Hunter's Hot Springs, Goose Lake, Warner Canyon, hanggliding[20] and areas for rock hunting.

Lake County is home to the Lakeview Hospital, and other regional health care facilities. It's also the location for The Warner Creek Correctional Facility opened in 2005 and is a 400-bed minimum security state prison. The prison employs an average of 110 correctional professionals and is located on a site comprising 91 acres (m^2). The facility itself occupies less than 15 acres (m^2) and utilizes approximately 117000 sq ft (m^2). of building space which includes areas for inmate housing, work and education programs, health services, food services, religious services, physical plant, warehouse and storage, vehicle maintenance, a laundry facility, recreational activities, administration and various other functions. A unique feature of the prison is its use of natural geothermal sources deep inside the Earth. The geothermally heated water is pumped up through a well and piped into a heat exchange unit where the heat is then transferred to the prison's water-loop system. Once the heat has been transferred and the water has cooled, the water is re-injected back into the ground.[21] Two prisoners briefly escaped in 2008.[22]

Politics

Though Lake County is located in central Oregon, politically it falls in line with the eastern side of the state. The majority of registered voters who are part of a political party in Lake County, as well as most counties in eastern Oregon, are members of the Republican Party.[23] In the 2008 presidential election, 71.53% of Lake County voters voted for Republican John McCain, while 25.95% voted for Democrat Barack Obama and 1.53% of voters either voted for a Third Party candidate or wrote in a candidate.[24] These numbers show a small shift towards the Democratic candidate when compared to the 2004 presidential election, in which 77.8% of Lake Country voters voted for George W. Bush, while 20.5% voted for John Kerry, and 1.7% of voters either voted for a Third Party candidate or wrote in a candidate.[25]

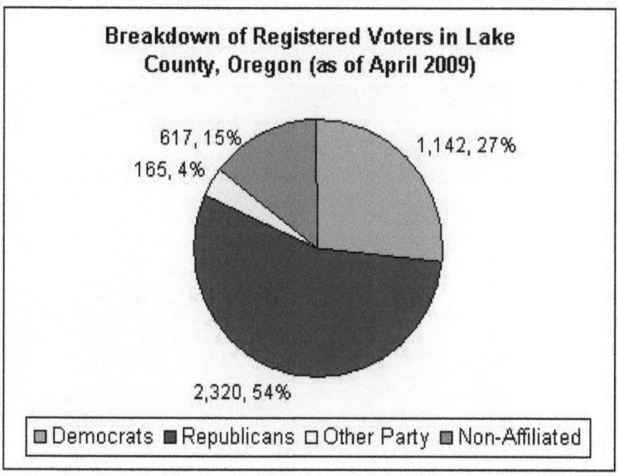

Geography

According to the U.S. Census Bureau, the county has a total area of 8358 square miles (km^2), of which 8136 square miles (km^2) is land and 223 square miles (km^2) (2.66%) is water.

Lake County ranges in elevation from 4130 to 8446 feet (to m).[16]

The area includes alkali lakes, high desert scrub, and forests of Western Juniper, Ponderosa Pine, and Lodgepole Pine in the hills and mountains. There are also several unusual geologic features and good areas for rockhounding and geological touring.

Lost Forest Research Natural Area is a protected relic forest closely associated with the Christmas Valley Sand Dunes in northeastern Lake County. Fossil Lake is an area nearby where many fossils of prehistoric animals have been identified. Glass Buttes is an obsidian complex in the extreme northeast corner of the county. Crack in the Ground, northeast of Christmas Valley is a long fissure with ice in its floor year round. Big Hole, Hole-in-the-Ground, and Fort Rock are ancient maar craters in the northwestern part of the county.

Watersheds

These twelve watersheds occur in whole or in part within Lake County:[26]

- Beaver-South Fork
- Goose Lake
- Guano
- Lake Abert
- Little Deschutes
- Lost
- Lower Crooked
- Silver
- Sprague
- Summer Lake
- Warner Lakes
- Williamson

Lakes

Named lakes wholly or partly in Lake County include:

- Abert Lake
- Alkali Lake[27]
- Campbell Lake
- Crump Lake
- Deadhorse Lake
- Dog Lake[28] [29]
- Drews Reservoir
- Goose Lake
- Hart Lake[30]
- Heart Lake [31]
- Silver Lake
- Summer Lake

The East shore of Lake Abert.

- Warner Lakes (Pelican, Crump, Hart, Anderson, Swamp, Mugwump, Flagstaff, Upper Campbell, Lower Campbell, Stone Corral, Turpin, and Bluejoint).[32] [33] [34] [35] [36]

Adjacent counties

- Deschutes County, Oregon - (north)
- Klamath County, Oregon - (west)
- Harney County, Oregon - (east)
- Modoc County, California - (south)
- Washoe County, Nevada - (south)

National protected areas

- Deschutes National Forest (part)
- Fremont National Forest (part)
- Hart Mountain National Antelope Refuge
- Sheldon National Wildlife Refuge (part)

Demographics

Historical populations		
Census	**Pop.**	**%±**
1880	2804	—
1890	2604	-7.1%
1900	2847	9.3%
1910	4658	63.6%
1920	3991	-14.3%
1930	4833	21.1%
1940	6293	30.2%
1950	6649	5.7%
1960	7158	7.7%
1970	6343	-11.4%
1980	7532	18.7%
1990	7186	-4.6%
2000	7422	3.3%
2010	7895	6.4%
[37] [38] [39]		

As of the census[40] of 2000, there were 7,422 people, 3,084 households, and 2,152 families residing in the county. The population density was 1 person per square mile (0/km^2). There were 3,999 housing units at an average density of 0 per square mile (0/km^2). The racial makeup of the county was 90.97% White, 0.13% Black or African American, 2.37% Native American, 0.71% Asian, 0.13% Pacific Islander, 3.19% from other races, and 2.48% from two or more races. 5.44% of the population were Hispanic or Latino of any race. 14.1% were of Irish, 14.0% United States or American, 13.8% German and 11.8% English ancestry according to Census 2000. 95.9% spoke English and 3.6% Spanish as their first language.

There were 3,084 households out of which 29.00% had children under the age of 18 living with them, 58.60% were married couples living together, 7.50% had a female householder with no husband present, and 30.20% were non-families. 26.20% of all households were made up of individuals and 11.10% had someone living alone who was 65 years of age or older. The average household size was 2.39 and the average family size was 2.84.

In the county, the population was spread out with 24.90% under the age of 18, 5.10% from 18 to 24, 24.30% from 25 to 44, 28.10% from 45 to 64, and 17.70% who were 65 years of age or older. The median age was 43 years. For every 100 females there were 100.50 males. For every 100 females age 18 and over, there were 98.30 males.

The median income for a household in the county was $29,506, and the median income for a family was $36,182. Males had a median income of $29,454 versus $23,475 for females. The per capita income for the county was $16,136. About 13.40% of families and 16.10% of the population were below the poverty line, including 20.40% of those under age 18 and 9.50% of those age 65 or over.

Arts and culture

Museums and other points of interest

North Lake County includes many geological sites including the Fort Rock, a crater marked by wave activity in what was once an ice age lake bed, and the Fort Rock Homestead Village Museum. It's also a site where camel bones and Native American artifacts have been uncovered. The "world's oldest shoes" were found here in 1938, changing the dates range scientists believed that humans inhabited the far west from 4,000 years ago.[41]

Several subsequent discoveries of even older sandals in the northern Great Basin confirmed the importance of archaeologist Luther Cressman's work. For this find, and for other research that broke down standing theories about the nature of the prehistoric Northwest, Cressman became known as the father of Oregon archaeology. [41] Other sights in North Lake include the Lost Forest, Crack-in-the-ground, and Hole-in-the-ground[42]

Oregon sunstones are found north of Plush.[43] Sunstone is Oregon's state gemstone.

Other areas of interest include, Abert Lake and Abert Rim, Goose Lake, Hunter's Hot Springs and its Old Perpetual Geyser, Schminck Memorial Museum and Lake County Museum,[44] Lake County Round-Up Museum, Schmink Museum, Warner Canyon ski area, Gearhart Mountain Wilderness, Sheldon National Wildlife Refuge, Summer Lake Hot Springs, sunstones (Oregon's state gemstone) near Plush, Warner Wetlands,[45] [46] Summer Lake Wildlife Area[47] and sections of the Fremont National Forest of the Fremont–Winema National Forests.

The Hart Mountain National Antelope Refuge includes big-horn sheep, pronghorn, sage grouse, predatory birds and migratory birds. The *Lake County Examiner* is located in Lakeview and has been published continuously since the late 19th century.[48]

Media

Newspapers

- *Desert Whispers*
- *Lake County Examiner*
- *North Lake County Outback News*

Infrastructure

Major highways

- Oregon Route 31
- U.S. Route 395

U.S. Highway 140

Passenger and freight railways

- Lake County Railroad

Communities

Incorporated cities

- Lakeview
- Paisley

Unincorporated communities

- Adel
- Blaisdell
- Christmas Valley
- Five Corners
- Fort Rock
- New Pine Creek
- Plush
- Quartz Mountain
- Silver Lake
- Summer Lake
- Valley Falls
- Westside

See also

- National Register of Historic Places listings in Lake County, Oregon

References

[1] http://www.lakecountyor.org

[2] McArthur, Lewis A. and Lewis L. McArthur, "Lake County", *Oregon Geographic Names* (Seventh Edition), Oregon Historical Society Press, Portland, Oregon, 2003, p. 553.

[3] Lake County, Oregon website (http://www.lakecountyor.org/)

[4] "Pre-Clovis Breakthrough" (http://www.archaeology.org/online/features/coprolites/). Archaeology.org. . Retrieved 2011-02-12.

[5] "Cultural Sequence in the Northern Great Basin: The View From Fort Rock" (http://darkwing.uoregon.edu/~anthro/academics/FieldSchool/section3.html). University of Oregon Department of Anthropology. . Retrieved 2007-04-09.

[6] "World's Oldest Shoes" (http://www.uoregon.edu/~connolly/FRsandals.htm). University of Oregon. . Retrieved 2007-04-09.

[7] "lakecountymuseum.com" (http://www.lakecountymuseum.com/Local History/timeline.html). lakecountymuseum.com. . Retrieved 2011-02-12.

[8] http://www.heraldandnews.com/articles/2008/10/12/news/diversions/doc48d2fbf09b840436604006.txt

[9] http://www.irishsheepherders.com/index.html

[10] "Oregon History Project" (http://www.ohs.org/education/oregonhistory/historical_records/dspDocument.cfm?doc_ID=8A032B61-99BB-85F8-059105E03E7F13D6). Ohs.org. . Retrieved 2011-02-12.

[11] "lakecountymuseum.com" (http://lakecountymuseum.com/). lakecountymuseum.com. . Retrieved 2011-02-12.
[12] Nevada-California-Oregon Railroad (http://www.trainweb.org/highdesertrails/nco.html) from trainweb.org
[13] "DOC Operations Division: Prison Warner Creek Correctional Facility" (http://www.oregon.gov/DOC/OPS/PRISON/wccf.shtml). Oregon.gov. 2010-03-16. . Retrieved 2011-02-12.
[14] Lake County Chamber of Commerce (http://www.lakecountychamber.org/facts/index.html)
[15] Collins Companies (http://www.collinswood.com/CertifiedForests/LakeviewSCS2.html)
[16] "Lakeview" (http://www.lakecountychamber.org/facts/index.html). Lakecountychamber.org. . Retrieved 2011-02-12.
[17] "Perlite Ore - We specialize in Perlite processing. Mining, and Marketing of Perlite" (http://www.cornerstonemineral.com/index.html). Cornerstonemineral.com. . Retrieved 2011-02-12.
[18] "Fremont Nat. Forest Uranium Mines (USDA)" (http://yosemite.epa.gov/r10/cleanup.nsf/d67b5aa82151d9ae88256da6005fb54e/8e2eef8a4cd8891b8825651a005885fa!OpenDocument). Yosemite.epa.gov. . Retrieved 2011-02-12.
[19] "White King/Lucky Lass" (http://www.fs.fed.us/r6/frewin/projects/forestplan/1990plan/freamend22rod.pdf) (PDF). . Retrieved 2011-02-12.
[20] Larry Leonard. "Oregon Magazine" (http://oregonmag.com/LakeCoTrav304.htm). Oregonmag.com. . Retrieved 2011-02-12.
[21] Warner Creek Correctional Facility (http://www.oregon.gov/DOC/OPS/PRISON/wccf.shtml)
[22] http://egov.oregon.gov/OSP/NEWSRL/news/09_07_2008_wccf_escape.shtml
[23] http://www.sos.state.or.us/elections/votreg/apr09.pdff Retrieved on 5/26/09
[24] "Dave Leip's Atlas of U.S. Presidential Elections - State Data" (http://www.uselectionatlas.org/RESULTS/statesub.php?year=2008&fips=41037&f=0&off=0&elect=0). Uselectionatlas.org. . Retrieved 2011-02-12.
[25] http://www.city-data.com/county/Lake_County-OR.html Retrieved on 5/26/09
[26] "Lake County | Surf Your Watershed | US EPA" (http://cfpub.epa.gov/surf/county.cfm?fips_code=41037). Cfpub.epa.gov. 2006-06-28. . Retrieved 2011-02-12.
[27] "Alkali Lake in Lake County Southern Oregon" (http://www.southernoregon.com/lakes/alkalilake/index.html). Southernoregon.com. . Retrieved 2011-02-12.
[28] "Fremont-Winema National Forests - Recreational Activities" (http://www.fs.fed.us/r6/frewin/recreation/campgrounds/developed/doglake/fre-doglk.shtml). Fs.fed.us. 2004-10-19. . Retrieved 2011-02-12.
[29] "Dog Lake in Fremont National Forest Southern Oregon" (http://www.southernoregon.com/lakes/doglake/index.html). Southernoregon.com. . Retrieved 2011-02-12.
[30] "Hart Lake Reservoir in Lake County Southern Oregon" (http://www.southernoregon.com/lakes/hartlakeresevoir/index.html). Southernoregon.com. . Retrieved 2011-02-12.
[31] "Heart Lake in Southern Oregon" (http://www.southernoregon.com/lakes/heartlake/index.html). Southernoregon.com. . Retrieved 2011-02-12.
[32] McArthur, Lewis A. and Lewis L. McArthur, "Warner Valley", *Oregon Geographic Names* (Seventh Edition), Oregon Historical Society Press, Portland, Oregon, 2003, p. 1010-1011.
[33] "Bluejoint Lake in Lake County Warner Valley Southern Oregon" (http://www.southernoregon.com/lakes/bluejointlake/index.html). Southernoregon.com. . Retrieved 2011-02-12.
[34] "Southern Oregon Lakes" (http://www.southernoregon.com/lakes/index.html). Southernoregon.com. . Retrieved 2011-02-12.
[35] Gottberg, John Anderson, "Pronghorn Paradise" (http://www.bendbulletin.com/apps/pbcs.dll/article?AID=/20090816/NEWS0107/908160416/1001/NEWS01&nav_category=NEWS01), *The Bulletin*, Bend, Oregon, 16 August 2009.
[36] Jackman, E.R. and R.A. Long, *The Oregon Desert* (http://books.google.com/books?id=U3ruodEof8MC&pg=PA361&lpg=PA361&dq="warner+valley"+oregon&source=bl&ots=SkQZkfYAif&sig=bZqcChbvOycOgO2R5ZNzgintnCU&hl=en&ei=yjHRSr-2Mo7ANuLAnZQD&sa=X&oi=book_result&ct=result&resnum=2&ved=0CBIQ6AEwAThG#v=onepage&q="warner valley" oregon&f=false), Canton Press, Caldwell, Idaho: 1964, p. 361.
[37] http://www.census.gov/population/www/censusdata/cencounts/files/or190090.txt
[38] http://factfinder2.census.gov
[39] http://mapserver.lib.virginia.edu/
[40] "American FactFinder" (http://factfinder.census.gov). United States Census Bureau. . Retrieved 2008-01-31.
[41] "Oregon Historical County Records Guide: Lake County History" (http://arcweb.sos.state.or.us/pages/records/local/county/lake/hist.html). Arcweb.sos.state.or.us. . Retrieved 2011-02-12.
[42] (http://www.christmasvalley.org/events.php)
[43] "About Lakeview | Oregon/Washington | Bureau of Land Management (BLM) | US Department of the Interior" (http://www.blm.gov/or/districts/lakeview/about.php). Blm.gov. . Retrieved 2011-02-12.
[44] "lakecountymuseum.com" (http://www.lakecountymuseum.com/). lakecountymuseum.com. . Retrieved 2011-02-12.
[45] "Warner Wetlands Details | Oregon/Washington | Bureau of Land Management (BLM) | US Department of the Interior" (http://www.blm.gov/or/resources/recreation/site_info.php?siteid=103). Blm.gov. . Retrieved 2011-02-12.
[46] "Warner Wetlands - OR" (http://www.paddling.net/places/showReport.html?1858). Paddling.net. 2007-06-25. . Retrieved 2011-02-12.
[47] "ODFW: Summer Lake Wildlife Area" (http://www.dfw.state.or.us/resources/visitors/summer_lake_wildlife_area.asp). Dfw.state.or.us. 2010-11-16. . Retrieved 2011-02-12.

[48] "Hart Mountain National Antelope Refuge" (http://www.fws.gov/sheldonhartmtn/Hart/index.html). Fws.gov. 2008-07-18. . Retrieved 2011-02-12.

External links

- Information on Lake County (http://bluebook.state.or.us/local/counties/counties19.htm) from the *Oregon Blue Book*
- Irish Sheepherders of Lake County Oregon (http://www.irishsheepherders.com/)

Deschutes_County,_Oregon

Deschutes County, Oregon	
Location in the state of Oregon	
Oregon's location in the U.S.	
Founded	December 13, 1916
Named for	Deschutes River
Seat	Bend
Largest city	Bend
Area **- Total** **- Land** **- Water**	3055 sq mi (7912 km²) 3018 sq mi (7816 km²) 37 sq mi (96 km²), 1.20%
Population **- (2010)** **- Density**	157733 52/sq mi (20.1/km²)
Time zone	Pacific: UTC-8/-7
Website	www.deschutes.org [1]

Deschutes County (pron.: /dəˈʃuːts/) is a county located in the U.S. state of Oregon. In 2010, its population was 157,733. The county was created in 1916 out of part of Crook County and was named for the Deschutes River, which itself was named by French-Canadian trappers of the early 19th century. It is part of the Bend, Oregon, Metropolitan Statistical Area and is the political and economic hub of Central Oregon. The county seat is Bend.

Deschutes is the fastest-growing county in Oregon.[2]

History

French-Canadian fur trappers of the Hudson's Bay Company gave the name Riviere des Chutes (River of the Falls) to the Deschutes River, from which the county derived its name.[3]

On December 13, 1916, Deschutes County was created from the southern part of Crook County. Bend has been the county seat since the county's formation. It was the last county in Oregon to be established.

The Shevlin-Hixon Lumber Company also operated within the Bend area processing Ponderosa pine trees.

Economy

During the 1990s, Deschutes County experienced the most rapid growth of any county in Oregon largely due to the availability of recreation activities year-round, and its location as the nearest population center to much of the central Cascade Range. Beyond tourism, principal industries in the county are lumber, ranching and agriculture—chiefly potatoes. The Forest Service owns 51% of the lands within the county boundaries.

Deschutes County is the home of four *destination resorts* as defined by Oregon's Department of Land Conservation and Development. These resort are major employers within the county. Three of these resorts, Sunriver, Eagle Crest, and Pronghorn, are among the county's ten largest tax payers.[4]

Politics

Deschutes County politically falls more in line with the eastern side of the state than the western side. The majority of registered voters who are part of a political party in Deschutes County, as well as most counties in eastern Oregon, are members of the Republican Party.[5] In the 2008 presidential election 48.96% of Deschutes County voters voted for Republican John McCain, while 48.66% voted for Democrat Barack Obama and 2.37% of voters either voted for a Third Party candidate or wrote in a candidate.[6] These numbers represented a significant shift towards the Democratic candidate when compared to the 2004 presidential election, in which 56.4% of Deschutes Country voters voted for George W. Bush, while 42.1% voted for John Kerry, and 1.5% of voters either voted for a third-party candidate or wrote in a candidate.[7]

In 2008, the Oregon house seat encompassing the city of Bend switched parties and hosted the only Democratic state legislator from a district east of the Cascades,[8] though the Republicans retook the seat in 2010.[9]

[5]

Geography

According to the U.S. Census Bureau, the county has a total area of 3055 square miles (7912.4 km^2), of which 3018 square miles (7816.6 km^2) is land and 37 square miles (95.8 km^2) is water. The total area is 1.20% water.

Adjacent counties

- Jefferson County - (north)
- Crook County - (east)
- Harney County (southeast)
- Klamath County - (south)
- Lake County - (south)
- Lane County - (west)
- Linn County - (northwest)

National protected areas

- Deschutes National Forest (part)
- Newberry National Volcanic Monument

Demographics

As of the census[10] of 2000, there were 115,367 people, 45,595 households, and 31,962 families residing in the county. The population density was 38 people per square mile (15/km²). There were 54,583 housing units at an average density of 18 per square mile (7/km²). The racial makeup of the county was 94.85% White, 0.19% Black or African American, 0.83% Native American, 0.74% Asian, 0.07% Pacific Islander, 1.36% from other races, and 1.96% from two or more races. 3.73% of the population were Hispanic or Latino of any race. 19.4% were of German, 13.2% English, 11.4% Irish and 9.1% American ancestry.

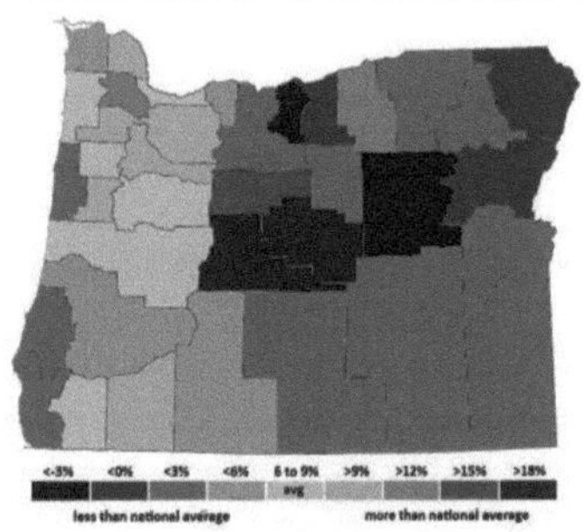

Deschutes county grew by 39.4% from 2000 to 2007, making it by far the fastest growing county in Oregon at more than four times the state average.

There were 45,595 households out of which 32.10% had children under the age of 18 living with them, 58.00% were married couples living together, 8.50% had a female householder with no husband present, and 29.90% were non-families. 22.00% of all households were made up of individuals and 7.70% had someone living alone who was 65 years of age or older. The average household size was 2.50 and the average family size was 2.91.

Historical populations		
Census	Pop.	%±
1920	9622	—
1930	14749	53.3%
1940	18631	26.3%
1950	21812	17.1%
1960	23100	5.9%
1970	30442	31.8%
1980	62142	104.1%
1990	74958	20.6%
2000	115367	53.9%
2010	157733	36.7%
sources:[11] [12]		

In the county, the population was spread out with 24.80% under the age of 18, 7.80% from 18 to 24, 28.60% from 25 to 44, 25.70% from 45 to 64, and 13.10% who were 65 years of age or older. The median age was 38 years. For every 100 females there were 98.70 males. For every 100 females age 18 and over, there were 97.00 males.

The median income for a household in the county was $41,847, and the median income for a family was $48,403. Males had a median income of $34,070 versus $25,069 for females. The per capita income for the county was $21,767. About 6.30% of families and 9.30% of the population were below the poverty line, including 10.40% of those under age 18 and 6.10% of those age 65 or over.

Geology

The overall defining landscape of Deschutes County is primarily marked with that of lava flows. Most of the lava found in Deschutes County issued forth from Newberry Volcano located just south of Bend. Because of this, many lava tubes are located within, including the prominent Lava River Cave. As a consequence, the county is the most cave-rich in the state of Oregon,[13] with over 250 known caves in the Deschutes National Forest alone.[14] Other prominent lava flows exist too, especially those related to the Mount Bachelor Volcanic Chain which consists of Mt. Bachelor as well as three smaller shield volcanos, and a series of cinder cones.[15]

Three Sisters mountains visible over a rail bridge crossing the Crooked River north of Terrebonne

Large volcanoes serve as a backdrop to the city of Bend in Deschutes County, ranging from the prominent Three Sisters, Mt. Washington, Mt. Bachelor, Broken Top, Newberry, Tumalo Mountain, Maiden Peak and others. A group of geologists have discovered Smith Rock State Park is part of an ancient supervolcano called the Crooked River caldera with a rim nearly six times the diameter of Newberry Volcano's caldera. This ancient supervolcano has long gone extinct. Its crater rim is barely recognizable as remnants mark portions of Powell Buttes, Gray Butte, and the western front of the Ochoco Mountains at Barnes Butte.[16]

On the eastern side of the county, it is mostly characterized by large buttes of much older volcanic origin. Most of these did not create any proper lava flows, or at least none that are known to exist. Some of the prominent buttes include: Horse Ridge, Pine Mountain, China Hat, and several others on the county border.

Two main types of lava flows are found within the county. The most common are the pāhoehoe flows which have been partially buried by volcanic ash, tephra deposits, and dirt over tens to hundreds of thousands of years. The ʻaʻā flows are fewer, but are much more prominent, with the most notable being associated with Lava Butte and the Lava Cast Forest in the Newberry National Volcanic Monument.

Communities

Incorporated cities

- Bend
- La Pine
- Redmond
- Sisters

View of the Cascades near La Pine, Oregon.

Unincorporated communities and CDPs

- Alfalfa
- Black Butte Ranch
- Brothers
- Cloverdale
- Deschutes
- Deschutes River Woods
- Elk Lake

- Hampton
- Millican
- Plainview
- Prineville Junction
- Pronghorn
- Seventh Mountain
- Shevlin
- Sunriver
- Terrebonne
- Tetherow
- Three Rivers
- Tumalo

See also

- National Register of Historic Places listings in Deschutes County, Oregon

References

[1] http://www.deschutes.org

[2] Bilby, Robert; Hanna, Susan; Huntly, Nancy; *et al.* (2007-07-08). "Human Population Impacts on Columbia River Basin Fish and Wildlife" (http://www.nwcouncil.org/library/isab/isab2007-3.pdf) (PDF). Independent Scientific Advisory Board. . Retrieved 2008-09-15.

[3] "Deschutes County" (http://bluebook.state.or.us/local/counties/counties09.htm). *Oregon Blue Book*. State of Oregon. . Retrieved 2009-04-10.

[4] Williams, Steve, "The Expansion of Oregon's Destination Resorts" (http://www.qualityinfo.org/olmisj/ArticleReader?itemid=00005396), WorkSource Oregon, *www.qualityinfo.org*, Oregon Employment Department, Salem, Oregon, 29 March 2007.

[5] (http://www.sos.state.or.us/elections/votreg/mar09.pdf)

[6] "Deschutes County Election Results 2009" (http://www.deschutes.org/Clerks-Office/Elections/Past-Elections/Results/20081104.aspx). Deschutes.org. . Retrieved 2009-04-22.

[7] "Deschutes County, Oregon (OR)" (http://www.city-data.com/county/Deschutes_County-OR.html). City-data.com. . Retrieved 2012-10-24.

[8] (http://judystiegler4hd54.com/wordpress/?p=32)

[9] "Jason Conger - Oregon State Representative, House District 54" (http://www.repjasonconger.com/). Repjasonconger.com. . Retrieved 2012-08-06.

[10] "American FactFinder" (http://factfinder.census.gov). United States Census Bureau. . Retrieved 2008-01-31.

[11] census.gov Oregon population by county, 1900-90 (http://www.census.gov/population/cencounts/or190090.txt) - accessed 2009-05-02

[12] "American FactFinder" (http://factfinder2.census.gov). Factfinder2.census.gov. . Retrieved 2012-08-06.

[13] "Oregon High Desert Grotto" (http://www.ohdgrotto.com). Ohdgrotto.com. . Retrieved 2012-10-24.

[14] Maben, Scott (March 11, 1994 p. A-17). "Guidelines Set to Save C.O. Caves" (http://news.google.com/newspapers?id=x2IPAAAAIBAJ&sjid=F4cDAAAAIBAJ&pg=6483,806941&dq=deschutes+cave&hl=en). *The Bulletin*. . Retrieved 2009-10-18.

[15] U.S. Department of the Interior; U.S. Geological Survey; Scott, William E.; Gardner, Cynthia A. (2009-10-29). *Geologic Map of the Mount Bachelor Volcanic Chain and Surrounding Area, Cascade Range, Oregon* (Map). 1 : 50,000. Cartography by D. F. Garcia; R. C. Ittner; S.E. Jefferies.

[16] McClaughry, Jason D.; Ferns, Mark L.;Gordon, Caroline L.;Patridge, Karyn A. (2009). "Field Trip Guide to the Oligocene Crooked River caldera: Central Oregon's Supervolcano, Crook, Deschutes, and Jefferson Counties, Oregon" (http://www.oregongeology.org/sub/publications/OG/OGv69n01.pdf). *Oregon Geology* (http://www.oregongeology.com/sub/quarpub/OrGeo.htm) **69** (1): 25–44. .

Oregon

State of Oregon	
 Flag (obverse) Seal	
Nickname(s): *Beaver State*	
Motto(s): *Alis volat propriis (Latin: She flies with her own wings)*	
Official language(s)	**De jure**: None[1] **De facto**: English
Demonym	Oregonian
Capital	Salem
Largest city	Portland
Largest metro area	Portland metropolitan area
Area	Ranked 9th in the U.S.
- Total	98,381 sq mi (255,026 km^2)
- Width	400 miles (640 km)
- Length	360 miles (580 km)
- % water	2.4
- Latitude	42° N to 46° 18′ N
- Longitude	116° 28′ W to 124° 38′ W
Population	Ranked 27th in the U.S.
- Total	3,899,353 (2012 est)[2]
- Density	39.9/sq mi (15.0/km^2) Ranked 39th in the U.S.
Elevation	
- Highest point	Mount Hood[3] [4] [5] 11,249 ft (3,428.8 m)
- Mean	3,300 ft (1,000 m)
- Lowest point	Pacific Ocean[4] sea level
Admission to Union	February 14, 1859 (33rd)
Governor	John Kitzhaber (D)
Secretary of State	Kate Brown (D)
Legislature	Legislative Assembly
- Upper house	State Senate

- Lower house	House of Representatives
U.S. Senators	Ron Wyden (D) Jeff Merkley (D)
U.S. House delegation	4 Democrats, 1 Republican (list)
Time zones	
- most of state	Pacific: UTC -8/-7
- most of Malheur County	Mountain: UTC -7/-6
Abbreviations	OR Ore. US-OR
Website	[www.oregon.gov www.oregon.gov]

Oregon (i/ˈɔɹɪɡən/ ***ORR**-ə-gən*)[7] is a state in the Pacific Northwest region of the United States. It is located on the Pacific coast, with Washington to the north, California to the south, Nevada on the southeast and Idaho to the east. The Columbia and Snake rivers delineate much of Oregon's northern and eastern boundaries, respectively. The area was inhabited by many indigenous tribes before the arrival of traders, explorers, and settlers who formed an autonomous government in Oregon Country in 1843. The Oregon Territory was created in 1848, and Oregon became the 33rd state on February 14, 1859.

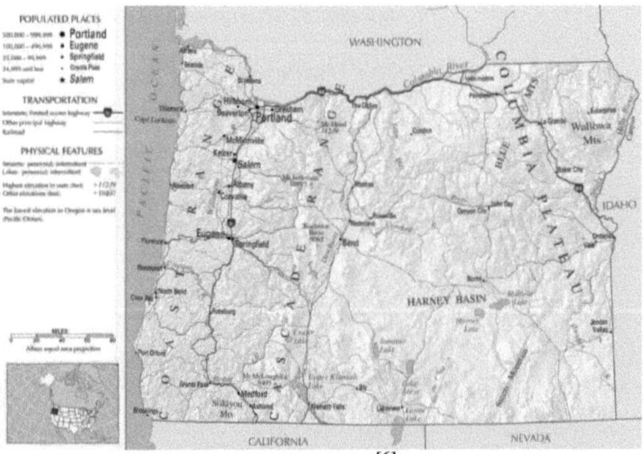

A map of Oregon.[6]

The Oregon coastline looking south from Ecola State Park, with Haystack Rock in the distance.

Oregon is the 9th most extensive and the 27th most populous of the 50 United States. Salem is the state's capital and third-most-populous city; Portland is the most populous. Portland is the 29th-largest U.S. city, with a population of 593,820 (2011 estimate) and a metro population of 2,262,605 (2011 estimate), the 23rd-largest U.S. metro area. The valley of the Willamette River in western Oregon is the state's most densely populated area and is home to eight of the ten most populous cities.

Oregon contains a diverse landscape including the windswept Pacific coastline, the volcanoes of the rugged and glaciated Cascade Mountain Range, many waterfalls (including Multnomah Falls), dense evergreen forests, mixed forests and deciduous forests at lower elevations, and high desert across much of the eastern portion of the state, extending into the Great Basin. The tall Douglas firs and redwoods along the rainy Western Oregon coast contrast with the lower density and fire-prone pine tree and juniper forests covering portions of the eastern half of the state. Alder trees are common in the west and fix nitrogen for the conifers; aspen groves are common in eastern Oregon.

Stretching east from Central Oregon, the state also includes semi-arid shrublands, prairies, deserts, steppes, and meadows. Mount Hood is the highest point in the state at 11249 feet (m). Crater Lake National Park is the only national park in Oregon.

Etymology

The earliest known use of the name, spelled **Ouragon**, was in a 1765 petition by Major Robert Rogers to the Kingdom of Great Britain. The term referred to the then–mythical River of the West (the Columbia River). By 1778 the spelling had shifted to *Oregon*.[8] In his 1765 petition, Rogers wrote:[9]

> "The rout [*sic*]...is from the Great Lakes towards the Head of the Mississippi, and from thence to the River called by the Indians Ouragon..."

One theory is the name comes from the French word *ouragan* ("windstorm" or "hurricane"), which was applied to the River of the West based on Native American tales of powerful Chinook winds of the lower Columbia River, or perhaps from firsthand French experience with the chinook winds of the Great Plains. At the time, the River of the West was thought to rise in western Minnesota and flow west through the Great Plains.[10]

Joaquin Miller explained in Sunset (magazine) in 1904 how Oregon's name was derived:[11]

> "The name, Oregon, is rounded down phonetically, from *Aure il agua*—Oragua, Or-a-gon, Oregon—given probably by the same Portuguese navigator that named the Farallones after his first officer, and it literally, in a large way, means cascades: 'Hear the waters.' You should steam up the Columbia and hear and feel the waters falling out of the clouds of Mount Hood to understand entirely the full meaning of the name *Aure il agua*, Oregon."

Another account, endorsed as the "most plausible explanation" in the book *Oregon Geographic Names*, was advanced by George R. Stewart in a 1944 article in *American Speech*. According to Stewart, the name came from an engraver's error in a French map published in the early 18th century, on which the Ouisiconsink (Wisconsin) River was spelled "Ouaricon-sint," broken on two lines with the -sint below, so there appeared to be a river flowing to the west named "Ouaricon."

According to the Oregon Tourism Commission (also known as Travel Oregon), present-day Oregonians /ˌɒrɪˈɡoʊniənz/[12] pronounce the state's name as "OR-UH-GUN, never OR-EE-GONE."[7]

After being drafted by the Detroit Lions in 2002, former Oregon Ducks quarterback Joey Harrington distributed "ORYGUN" stickers to members of the media as a reminder of how to pronounce the name of his home state.[13] [14] The stickers are sold by the University of Oregon Bookstore, which credits the spelling as a joke that is meant "for Oregonians and Oregon fans everywhere who get a kick out of this hilarious mispronunciation of our state."[15]

History

Humans have inhabited the area that is now Oregon for at least 15,000 years. In recorded history, mentions of the land date to as early as the 16th century. During the 18th and 19th centuries, European powers – and later the United States – quarreled over possession of the region until 1846 when the U.S. and Great Britain finalized division of the region. Oregon became a state in 1859 and is now home to over 3.8 million residents.

Earliest inhabitants

Human habitation of the Pacific Northwest began at least 15,000 years ago, with the oldest evidence of habitation in Oregon found at Fort Rock Cave and the Paisley Caves in Lake County. Archaeologist Luther Cressman dated material from Fort Rock to 13,200 years ago.[16] By 8000 B.C. there were settlements throughout the state, with populations concentrated along the lower Columbia River, in the western valleys, and around coastal estuaries.

By the 16th century, Oregon was home to many Native American groups, including the Coquille (Ko-Kwell), Bannock, Chasta, Chinook, Kalapuya, Klamath, Molalla, Nez Perce, Takelma, and Umpqua.[17] [18] [19] [20]

European exploration

The first Europeans to visit Oregon were Spanish explorers led by Juan Rodríguez Cabrillo who sighted southern Oregon off the Pacific Coast in 1543. In 1592, Juan de Fuca undertook detailed mapping and ocean current studies. Stops along these trips included Oregon as well as the strait now bearing his name and the future emplacement of Vancouver. Exploration was retaken routinely in 1774, starting by the expedition of frigate *Santiago* by Juan José Pérez Hernández (see Spanish expeditions to the Pacific Northwest), and the coast of Oregon became a valuable trading route to Asia. In 1778, British captain James Cook also explored the coast.[21] French Canadian and metis trappers and missionaries arrived in the eastern part of the state in the late 18th century and early 19th century, many having travelled as members of Lewis and Clark and the 1811 Astor expeditions.

Some stayed permanently, including Étienne Lussier, believed to be the first European farmer in the state of Oregon. The evidence of this French Canadian presence can be found in the numerous names of French origin in that part of the state: Charbonneau, Malheur Lake and River, Grande Ronde and Des Chutes Rivers, cities of La Grande, Ontario, etc.

During U.S. westward expansion

The Lewis and Clark Expedition traveled through the region also in search of the Northwest Passage. They built their winter fort in 1805-06 at Fort Clatsop, near the mouth of the Columbia River. British explorer David Thompson also conducted overland exploration.

In 1811, David Thompson, of the North West Company, became the first European to navigate the entire Columbia River. Stopping on the way, at the junction of the Snake River, he posted a claim to the region for Great Britain and the North West Company. Upon returning to Montreal, he publicized the abundance of fur-bearing animals in the area.

Also in 1811, New Yorker John Jacob Astor financed the establishment of Fort Astoria at the mouth of the Columbia River as a western outpost to his Pacific Fur Company;[22] this was the first permanent European settlement in Oregon.

In the War of 1812, the British gained control of all Pacific Fur Company posts. The Treaty of 1818 established joint British and American occupancy of the region west of the Rocky Mountains to the Pacific Ocean. By the 1820s and 1830s, the Hudson's Bay Company dominated the Pacific Northwest from its Columbia District headquarters at Fort Vancouver (built in 1825 by the District's Chief Factor John McLoughlin across the Columbia from present-day Portland).

In 1841, the expert trapper and entrepreneur Ewing Young died leaving considerable wealth and no apparent heir, and no system to probate his estate. A meeting followed Young's funeral at which a probate government was proposed. Doctor Ira Babcock of Jason Lee's Methodist Mission was elected Supreme Judge. Babcock chaired two meetings in 1842 at Champoeg, (half way between Lee's mission and Oregon City), to discuss wolves and other animals of contemporary concern. These meetings were precursors to an all-citizen meeting in 1843, which instituted a provisional government headed by an executive committee made up of David Hill, Alanson Beers, and Joseph Gale. This government was the first acting public government of the Oregon Country before annexation by the government of the United States.

Also in 1841, Sir George Simpson, Governor of the Hudson's Bay Company, reversed the Hudson's Bay Company's long-standing policy of discouraging settlement because it interfered with the lucrative fur trade. He directed that some 200 Red River Colony settlers be relocated to HBC farms near Fort Vancouver, (the James Sinclair expedition), in an attempt to hold Columbia District.

Starting in 1842–1843, the Oregon Trail brought many new American settlers to Oregon Country. For some time, it seemed that Britain and the United States would go to war for a third time in 75 years (see Oregon boundary dispute), but the border was defined peacefully in 1846 by the Oregon Treaty. The border between the United States and British North America was set at the 49th parallel. The Oregon Territory was officially organized in 1848.

Settlement increased with the Donation Land Claim Act of 1850 and the forced relocation of the native population to Indian reservations in Oregon.

After statehood

Oregon was admitted to the Union on February 14, 1859. Founded as a refuge from disputes over slavery, Oregon had a "whites only" clause in its original state Constitution.[23]

Panorama of Portland, Oregon, in 1890.

At the outbreak of the American Civil War, regular U.S. troops were withdrawn and sent east. Volunteer cavalry recruited in California were sent north to Oregon to keep peace and protect the populace. The First Oregon Cavalry served until June 1865.

In the 1880s, the growth of railroads helped market the state's lumber, wheat, and the rapid growth of its cities.

20th and 21st centuries

In 1902, Oregon introduced direct legislation by the state's citizens through initiatives and referenda, known as the Oregon System. Oregon state ballots often include politically conservative proposals side-by-side with politically liberal ones, illustrating the diversity of political thought in the state.

Industrial expansion began in earnest following the 1933–1937 construction of the Bonneville Dam on the Columbia River. Hydroelectric power, food, and lumber provided by Oregon helped fuel the development of the West, although the periodic fluctuations in the U.S. building industry have hurt the state's economy on multiple occasions.

Geography

National parks and historic areas in Oregon

Entity	Location
Crater Lake National Park	Southern Oregon
John Day Fossil Beds National Monument	Eastern Oregon
Newberry National Volcanic Monument	Central Oregon
Cascade–Siskiyou National Monument	Southern Oregon
Oregon Caves National Monument	Southern Oregon
California Trail	Southern Oregon, California
Fort Vancouver National Historic Site	Western Oregon, Washington
Lewis and Clark National Historic Trail	IL, MO, KS, IA, NE, SD, ND, MT, ID, **OR**, WA
Lewis and Clark National and State Historical Parks	Western Oregon, Washington
Nez Perce National Historical Park	MT, ID, **OR**, WA

Oregon Trail	MO, KS, NE, WY, ID, **OR**

Oregon's geography may be split roughly into eight areas:

- Oregon Coast—west of the Coast Range
- Willamette Valley
- Rogue Valley
- Cascade Mountains
- Klamath Mountains
- Columbia River Plateau
- Oregon Outback
- Blue Mountains (ecoregion)

The mountainous regions of western Oregon, home to three of the most prominent mountain peaks of the United States including Mount Hood, were formed by the volcanic activity of the Juan de Fuca Plate, a tectonic plate that poses a continued threat of volcanic activity and earthquakes in the region. The most recent major activity was the 1700 Cascadia earthquake. (Washington's Mount St. Helens erupted in 1980, an event which was visible from Oregon.)

The Columbia River, which forms much of the northern border of Oregon, also played a major role in the region's geological evolution, as well as its economic and cultural development. The Columbia is one of North America's largest rivers, and one of two rivers to cut through the Cascades (the Klamath River in Southern Oregon is the other). About 15,000 years ago, the Columbia repeatedly flooded much of Oregon during the Missoula Floods; the modern fertility of the Willamette Valley is largely a result of those floods. Plentiful salmon made parts of the river, such as Celilo Falls, hubs of economic activity for thousands of years. In the 20th century, numerous hydroelectric dams were constructed along the Columbia, with major impacts on salmon, transportation and commerce, electric power, and flood control.

Today, Oregon's landscape varies from rain forest in the Coast Range to barren desert in the southeast, which still meets the technical definition of a frontier.

Oregon is 295 miles (475 km) north to south at longest distance, and 395 miles (636 km) east to west at longest distance. In land and water area, Oregon is the ninth largest state, covering 98381 square miles (km^2).[24] The highest point in Oregon is the summit of Mount Hood, at 11239 feet (3426 m), and its lowest point is sea level of the Pacific Ocean along the Oregon coast.[25] Its mean elevation is 3300 feet (1006 m). Crater Lake National Park is the state's only national park and the site of Crater Lake, the deepest lake in the U.S. at 1943 feet (592 m).[26] Oregon claims the D River is the shortest river in the world,[27] though the American state of Montana makes the same claim of its Roe River.[28]

Oregon is also home to Mill Ends Park (in Portland),[29] the smallest park in the world at 452 square inches (0.29 m^2). Oregon's geographical center is farther west than that of any of the other 48 contiguous states (although the westernmost point of the lower 48 states is in Washington). Its antipodes, diametrically opposite its geographical center on the Earth's surface, is at 44°00′S 59°30′E in the Indian Ocean northwest of Port-aux-Français in the French Southern and Antarctic Lands. Oregon lies in two time zones. Most of Malheur County is in the Mountain Time Zone while the rest of the state lies in the Pacific Time Zone.

Oregon is home to what is considered the largest single organism in the world, an *Armillaria solidipes* fungus beneath the Malheur National Forest of eastern Oregon.[30]

Images of Oregon

Mount Hood, with Trillium Lake in the foreground

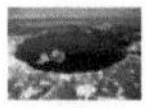

An aerial view of Crater Lake in Oregon

Sunset over Malheur Butte, an extinct volcanic cinder cone near Ontario, Oregon.

Downtown Eugene as seen from Skinner Butte in North Eugene

Nearly half of Oregon's land is held by the U.S. Forest Service and the Bureau of Land Management.[31]

Willamette Valley basin map

Major cities

Most Populous Cities

City	Population (2010 US Census)[32]
1. Portland	583,776
2. Eugene	156,185
3. Salem	154,637
4. Gresham	105,594
5. Hillsboro	91,611
6. Beaverton	89,803
7. Bend	76,639
8. Medford	74,907
9. Springfield	59,403
10. Corvallis	54,462
11. Albany	50,158
12. Tigard	48,035

Oregon's population is largely concentrated in the Willamette Valley, which stretches from Eugene in the south (home of the University of Oregon) through Corvallis (home of Oregon State University) and Salem (the capital) to Portland (Oregon's largest city).[32]

Astoria, at the mouth of the Columbia River, was the first permanent English-speaking settlement west of the Rockies in what is now the United States. Oregon City, at the end of the Oregon Trail, was the Oregon Territory's first incorporated city, and was its first capital from 1848 until 1852, when the capital was moved to Salem. Bend, near the geographic center of the state, is one of the ten fastest-growing metropolitan areas in the United States.[33] In the southern part of the state, Medford is a rapidly growing metro area, which is home to The Rogue Valley International-Medford Airport, the third-busiest airport in the state. To the south, near the California-Oregon border, is the community of Ashland, home of the Tony Award-winning Oregon Shakespeare Festival.

Climate

Oregon's climate – particularly in the western part of the state – is heavily influenced by the Pacific Ocean. The climate is mild, but periods of extreme hot and cold can affect parts of the state. Oregon's population centers, which lie mostly in the western part of the state, are moist and mild, while the lightly populated high deserts of Central and Eastern Oregon are much drier. Oregon's highest recorded temperature is 119 °F (°C) at Pendleton on August 10, 1898, and the lowest recorded temperature is –54 °F (– °C) at Seneca on February 10, 1933.

Governance

A writer in the Oregon Country book *A Pacific Republic*, written in 1839, predicted the territory was to become an independent republic. Four years later, in 1843, settlers of the Willamette Valley voted in majority for a republic government.[34] The Oregon Country functioned in this way until August 13, 1848, when Oregon was annexed by the United States and a territorial government was established. Oregon maintained a territorial government until February 14, 1859, when it was granted statehood.[35]

State

Oregon state government has a separation of powers similar to the federal government. It has three branches, called departments by the state's constitution:

- a legislative department (the bicameral Oregon Legislative Assembly),
- an executive department which includes an "administrative department" and Oregon's governor serving as chief executive, and
- a judicial department, headed by the Chief Justice of the Oregon Supreme Court.

Governors in Oregon serve four-year terms and are limited to two consecutive terms, but an unlimited number of total terms. Oregon has no lieutenant governor; in the event that the office of governor is vacated, Article V, Section 8a of the Oregon Constitution specifies that the Secretary of State is first in line for succession.[36] The other statewide officers are Treasurer, Attorney General, Superintendent, and Labor Commissioner. The biennial Oregon Legislative Assembly consists of a thirty-member Senate and a sixty-member House. The state supreme court has seven elected justices, currently including the only two openly gay state supreme court justices in the nation. They choose one of their own to serve a six-year term as Chief Justice. The only court that may reverse or modify a decision of the Oregon Supreme Court is the Supreme Court of the United States.

The debate over whether to move to annual sessions is a long-standing battle in Oregon politics, but the voters have resisted the move from citizen legislators to professional lawmakers. Because Oregon's state budget is written in two-year increments and, having no sales tax, its revenue is based largely on income taxes, it is often significantly over- or under-budget. Recent legislatures have had to be called into special session repeatedly to address revenue shortfalls resulting from economic downturns, bringing to a head the need for more frequent legislative sessions. Oregon Initiative 71, passed in 2010, mandates the Legislature to begin meeting every year, for 160 days in odd-numbered years, and 35 days in even-numbered years.

Oregon State Capitol.

The state maintains formal relationships with the nine federally recognized tribes in Oregon:

- Burns Paiute Tribe
- Confederated Tribes of Coos, Lower Umpqua and Siuslaw Indians
- Confederated Tribes of Grand Ronde
- Confederated Tribes of Siletz Indians
- Confederated Tribes of Warm Springs
- Confederated Tribes of the Umatilla Indian Reservation
- Cow Creek Band of Umpqua Tribe of Indians
- Klamath Tribes
- Coquille Indian Tribe

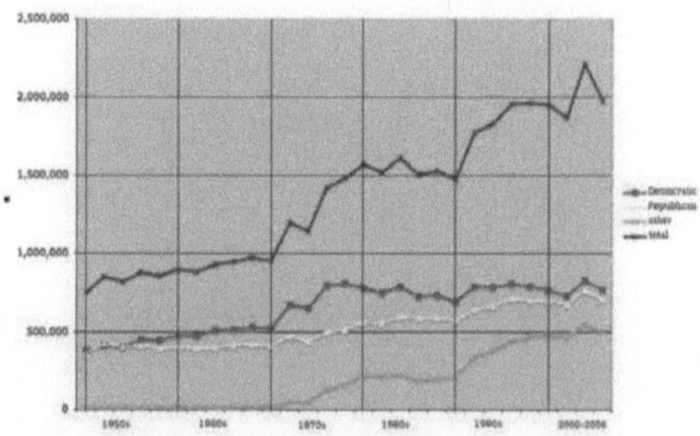

Party registration in Oregon, 1950–2006. total Democratic Party Republican Party non-affiliated and minor parties

Oregonians have voted for the Democratic Presidential candidate in every election since 1988. In 2004 and 2006, Democrats won control of the state Senate and then the House. Since the late 1990s, Oregon has been represented by four Democrats and one Republican in the U.S. House of Representatives. Since 2009, the state has had two Democratic Senators, Ron Wyden and Jeff Merkley. Oregon voters have elected Democratic governors in every election since 1986, most recently electing John Kitzhaber over Republican Chris Dudley in 2010.

The base of Democratic support is largely concentrated in the urban centers of the Willamette Valley. The eastern two-thirds of the state beyond the Cascade Mountains typically votes Republican; in 2000 and 2004, George W. Bush carried every county east of the Cascades. However, the region's sparse population means that the more populous counties in the Willamette Valley usually outweigh the eastern counties in statewide elections.

Oregon's politics are largely similar to those of neighboring Washington – for instance, in the contrast between urban and rural issues.

In the 2002 general election, Oregon voters approved a ballot measure to increase the state minimum wage automatically each year according to inflationary changes, which are measured by the consumer price index (CPI).[37] In the 2004 general election, Oregon voters passed ballot measures banning same-sex marriage,[38] and restricting land use regulation.[39] In the 2006 general election, voters restricted the use of eminent domain and extended the state's discount prescription drug coverage.[40]

The distribution, sales and consumption of alcoholic beverages are regulated in the state by the Oregon Liquor Control Commission. Thus, Oregon is an Alcoholic beverage control state. While wine and beer are available in most grocery stores, few stores sell hard liquor.

In March 2011, Oregon ranked amongst the top seven "Best" states in the American State Litter Scorecard, for overall effectiveness and quality of its public space cleanliness—-primarily roadway and adjacent litter—from state and related debris removal efforts.[41]

Federal

Like all US states, Oregon is represented by two U.S. Senators. Since the 1980 census, Oregon has had five Congressional districts.

After Oregon was admitted to the Union, it began with a single member in the House of Representatives (La Fayette Grover, who served in the 35th United States Congress for less than a month). Congressional apportionment increased the size of the delegation following the censuses of 1890, 1910, 1940, and 1980. A detailed list of the past and present Congressional delegations from Oregon is available.

The United States District Court for the District of Oregon hears federal cases in the state. The court has courthouses in Portland, Eugene, Medford, and Pendleton. Also in Portland is the federal bankruptcy court, with a second branch in Eugene.[42] Oregon (among other western states and territories) is in the 9th Court of Appeals. One of the court's meeting places is at the Pioneer Courthouse in downtown Portland, a National Historic Landmark built in 1869.

Politics

Presidential elections results[43]

Year	Republican	Democratic
2012	42.18% *754,095*	**54.27%** *970,343*
2008	40.40% *738,475*	**56.75%** *1,037,291*
2004	47.19% *866,831*	**51.35%** *943,163*
2000	46.46% *713,577*	**47.01%** *720,342*
1996	39.06% *538,152*	**47.15%** *649,641*
1992	32.53% *475,757*	**42.48%** *621,314*
1988	46.61% *560,126*	**51.28%** *616,206*
1984	**55.91%** *685,700*	43.74% *536,479*
1980	**48.33%** *571,044*	38.67% *456,890*
1976	**47.78%** *492,120*	47.62% *490,407*
1972	**52.45%** *486,686*	42.33% *392,760*
1968	**49.83%** *408,433*	43.78% *358,866*
1964	35.96% *282,779*	**63.72%** *501,017*
1960	**52.56%** *408,060*	47.32% *367,402*
1956	**55.25%** *406,393*	44.75% *329,204*
1952	**60.54%** *420,815*	38.93% *270,579*

The state has been thought of as politically split by the Cascade Range, with western Oregon being liberal and Eastern Oregon being conservative. In a 2008 analysis of the 2004 presidential election, a political analyst found that according to the application of a Likert scale, Oregon boasted both the most liberal Kerry voters and the most conservative Bush voters, making it the most politically polarized state in the country.[44]

During Oregon's history it has adopted many electoral reforms proposed during the Progressive Era, through the efforts of William S. U'Ren and his Direct Legislation League. Under his leadership, the state overwhelmingly approved a ballot measure in 1902 that created the initiative and referendum for citizens to introduce or approve proposed laws or amendments to the state constitution directly, making Oregon the first state to adopt such a system. Today, roughly half of U.S. states do so.[45]

In following years, the primary election to select party candidates was adopted in 1904, and in 1908 the Oregon Constitution was amended to include recall of public officials. More recent amendments include the nation's first doctor-assisted suicide law,[46] called the Death with Dignity law (which was challenged, unsuccessfully, in 2005 by the Bush administration in a case heard by the U.S. Supreme Court), legalization of medical cannabis, and among the nation's strongest anti-urban sprawl and pro-environment laws. More recently, 2004's Measure 37 reflects a backlash against such land-use laws. However, a further ballot measure in 2007, Measure 49, curtailed many of the provisions of 37.

Of the measures placed on the ballot since 1902, the people have passed 99 of the 288 initiatives and 25 of the 61 referendums on the ballot, though not all of them survived challenges in courts (see *Pierce v. Society of Sisters*, for an example). During the same period, the legislature has referred 363 measures to the people, of which 206 have passed.

Oregon pioneered the American use of postal voting, beginning with experimentation approved by the Oregon Legislative Assembly in 1981 and culminating with a 1998 ballot measure mandating that all counties conduct elections by mail. It remains the only state, with the exception of Washington, where voting by mail is the only method of voting.[47]

Under the leadership of Governor John Kitzhaber in 1994, Oregon was the first state in the US to set up effective health care programs with the Oregon Health Plan, which made health care available to most of its citizens without private health insurance.

In the U.S. Electoral College, Oregon casts seven votes. Oregon has supported Democratic candidates in the last six elections. Democrat Barack Obama won the state in 2008 by a margin of sixteen percentage points, with over 56% of the popular vote.

Economy

The Gross Domestic Product (GDP) of Oregon in 2010 was $168.6 billion, it is the United States's 26th wealthiest state by GDP. The state's per capita personal income in 2010 was $44,447.[48]

A grain elevator in Halsey storing grass seed, one of the state's largest crops.

Agriculture

Land in the Willamette Valley owes its fertility to the Missoula Floods, which deposited lake sediment from Glacial Lake Missoula in western Montana onto the valley floor.[49]

Oregon is also one of four major world hazelnut growing regions, and produces 95% of the domestic hazelnuts in the United States. While the history of the wine production in Oregon can be traced to before Prohibition, it became a significant industry beginning in the 1970s. In 2005, Oregon ranked third among U.S. states with 303 wineries.[50] Due to regional similarities in climate and soil, the grapes planted in Oregon are often the same varieties found in the French regions of Alsace and Burgundy.

In the Southern Oregon coast commercially cultivated cranberries account for about 7 percent of U.S. production, and the cranberry ranks twenty-third among Oregon's top fifty agricultural commodities. From 2006 to 2008, Oregon growers harvested between forty and forty-nine million pounds of berries every year. Cranberry cultivation in Oregon uses about 27,000 acres in southern Coos and northern Curry counties, centered around the coastal city of Bandon, Oregon.

In the northeastern region of the state, particularly around Pendleton, both irrigated and dry land wheat is grown. Oregon farmers and ranchers also produce cattle, sheep, dairy products, eggs and poultry.

Forestry and fisheries

Vast forests have historically made Oregon one of the nation's major timber production and logging states, but forest fires (such as the Tillamook Burn), over-harvesting, and lawsuits over the proper management of the extensive federal forest holdings have reduced the timber produced. According to the Oregon Forest Resources Institute, between 1989 and 2001 the amount of timber harvested from federal lands dropped some 96%, from 4,333 million to 173 million board feet (10,000,000 to 408,000 m^3), although harvest levels on private land have remained relatively constant.[51]

Even the shift in recent years towards finished goods such as paper and building materials has not slowed the decline of the timber industry in the state. The effects of this decline have included Weyerhaeuser's acquisition of Portland-based Willamette Industries in January 2002, the relocation of Louisiana-Pacific's corporate headquarters from Portland to Nashville, and the decline of former lumber company towns such as Gilchrist. Despite these changes, Oregon still leads the United States in softwood lumber production; in 2001, 6,056 million board feet (14,000,000 m^3) was produced in Oregon, compared with 4,257 million board feet (10,050,000 m^3) in Washington, 2,731 million board feet (6,444,000 m^3) in California, 2,413 million board feet (5,694,000 m^3) in Georgia, and 2,327 million board feet (5,491,000 m^3) in Mississippi.[52] The slow of the timber and lumber industry has caused high unemployment rates in rural areas.[53]

Oregon has one of the largest salmon-fishing industries in the world, although ocean fisheries have reduced the river fisheries in recent years. See also the List of freshwater fishes of Oregon.

Tourism

Tourism is also a strong industry in the state. Oregon's mountains, forests, waterfalls, lakes (including Crater Lake National Park), and beaches draw visitors year round. The Oregon Shakespeare Festival, held in Ashland, is a tourist draw for Southern Oregon.

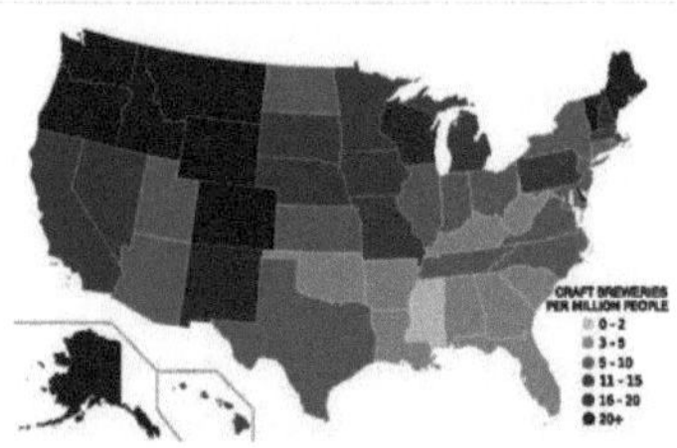

Oregon ranks 4th nationally in craft breweries per capita.

Oregon is home to many breweries and Portland has the largest number of breweries of any city in the world.[54]

Oregon occasionally hosts film shoots. Movies filmed in Oregon include: *Animal House*, *Free Willy*, *The General*, *The Goonies*, *Kindergarten Cop*, *One Flew Over the Cuckoo's Nest*, and *Stand By Me*. Oregon native Matt Groening, creator of *The Simpsons*, has incorporated many references from his hometown of Portland into the TV series.[55]

Technology

High technology industries and services have been a major employer since the 1970s. Tektronix was the largest private employer in Oregon until the late 1980s. Intel's creation and expansion of several facilities in eastern Washington County continued the growth that Tektronix had started. Intel, the state's largest for-profit private employer, operates four large facilities, with Ronler Acres, Jones Farm and Hawthorn Farm all located in Hillsboro.[56]

The spinoffs and startups that were produced by these two companies led to the establishment in that area of the so-called Silicon Forest. The recession and dot-com bust of 2001 hit the region hard; many high technology employers reduced the number of their employees or went out of business. Open Source Development Labs made news in 2004 when they hired Linus Torvalds, developer of the Linux kernel. Recently, biotechnology giant Genentech purchased several acres of land in Hillsboro to expand its production capabilities.[57] Oregon is home to several large datacenters that take advantage of cheap power and a climate in Central Oregon conducive to reducing cooling costs. Google has a large datacenter in The Dalles; Facebook has built a large datacenter in Prineville; and Amazon is restarting construction of a datacenter in Boardman.

Corporate headquarters

Largest Public Corporations Headquartered in Oregon[58]

Corporation	Headquarters	Market cap (million)
1. Nike, Inc.	near Beaverton	$32,039
2. Precision Castparts Corp.	Portland	$16,158
3. FLIR Systems	Wilsonville	$4,250
4. StanCorp Financial Group	Portland	$2,495
5. Schnitzer Steel Industries	Portland	$1,974
6. Portland General Electric	Portland	$1,737
7. Columbia Sportswear	near Beaverton	$1,593
8. Northwest Natural Gas	Portland	$1,287
9. Mentor Graphics	Wilsonville	$976
10. TriQuint Semiconductor	Hillsboro	$938

Oregon is also the home of large corporations in other industries. The world headquarters of Nike, Inc. are located near Beaverton. Medford is home to Harry and David, which sells gift items under several brands. Medford is also home to the national headquarters of the Fortune 1000 company, Lithia Motors. Portland is home to one of the West's largest trade book publishing houses, Graphic Arts Center Publishing. Oregon is also home to Mentor Graphics Corporation, a world leader in electronic design automation (EDA) located in Wilsonville and employs roughly 4,500 people worldwide.

Employment

As of September 2012, the state's unemployment rate is 8.7%.[59] Oregon's largest for-profit employer is Intel, located in the Silicon Forest area on Portland's west side. Intel was the largest employer in Oregon until 2008. As of January 2009, the largest employer in Oregon is Providence Health & Services, a non-profit.[60]

Taxes and budgets

Oregon's biennial state budget, $42.4 billion as of 2007, comprises General Funds, Federal Funds, Lottery Funds, and Other Funds. Personal income taxes account for 88% of the General Fund's projected funds.[61] The Lottery Fund, which has grown steadily since the lottery was approved in 1984, exceeded expectations in the 2007 fiscal years, at $604 million.[62]

Oregon is one of only five states that have no sales tax.[63] Oregon voters have been resolute in their opposition to a sales tax, voting proposals down each of the nine times they have been presented.[64] The last vote, for 1993's Measure 1, was defeated by a 72–24% margin.[65]

The state also has a minimum corporate tax of only $10 a year, amounting to 5.6% of the General Fund in the 2005–2007 biennium; data about which businesses pay the minimum is not available to the public.[66] As a result, the state relies on property and income taxes for its revenue. Oregon has the fifth highest personal income tax in the nation. According to the U.S. Census Bureau, Oregon ranked 41st out of the 50 states in taxes per capita in 2005.[67] The average paid of $1,791.45 is higher than only nine other states.[67]

Some local governments levy sales taxes on services: the city of Ashland, for example, collects a 5% sales tax on prepared food.[68]

Oregon is one of six states with a revenue limit.[69] The "kicker law" stipulates that when income tax collections exceed state economists' estimates by 2% or more, any excess must be returned to taxpayers.[70] Since the enactment of the law in 1979, refunds have been issued for seven of the eleven biennia.[71] In 2000, Ballot Measure 86 converted the "kicker" law from statute to the Oregon Constitution, and changed some of its provisions.

Federal payments to county governments, which were granted to replace timber revenue when logging in National Forests was restricted in the 1990s, have been under threat of suspension for several years. This issue dominates the future revenue of rural counties, which have come to rely on the payments in providing essential services.[72]

55 percent of state revenues are spent on public education, 23% on human services (child protective services, Medicaid, and senior services), 17% on public safety, and 5% on other services.[73]

Demographics

Population

Historical populations			
Census	Pop.		%±
1850	12093		—
1860	52465		333.8%
1870	90923		73.3%
1880	174768		92.2%
1890	317704		81.8%
1900	413536		30.2%
1910	672765		62.7%

1920	783389	16.4%
1930	953786	21.8%
1940	1089684	14.2%
1950	1521341	39.6%
1960	1768687	16.3%
1970	2091533	18.3%
1980	2633156	25.9%
1990	2842321	7.9%
2000	3421399	20.4%
2010	3831074	12.0%
Est. 2012	3899353	1.8%
U.S. Census Bureau[74]		

The United States Census Bureau estimates that the population of Oregon was 3,899,353 on July 1, 2012, a 1.8% increase since the 2010 United States Census.[2]

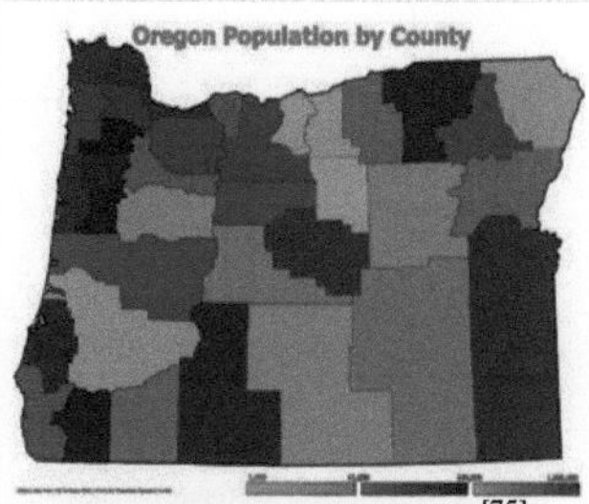

Source: Population Research Center[75]

As of the census of 2010,[76] Oregon has a population of 3,831,074, which is an increase of 409,675, or 12%, since the year 2000. The population density is 39.9 persons per square mile. There are 1,675,562 housing units, a 15.3% increase over 2000. Among them, 90.7% are occupied.

Hispanics or Latinos make up 11.7% of the total population. Among those who are not Hispanic or Latino, 78.5% is "white alone," 1.7% is "black or African American alone," 1.1% is "American Indian or Alaska native alone," 3.6% is "Asian alone," 0.3% is "Native Hawaiian and other Pacific Islander alone," 0.1% is "another race alone," and 2.9% is multiracial.

The state's most populous ethnic group, non-Hispanic white, has declined from 95.8% in 1970 to 78.1% in 2011.[77][78] As of 2011, 38.7% of Oregon's children under the age of 1 belonged to minority groups.[79]

Of the state's total population, 22.6% was under age 18, and 77.4% were 18 or older.

The center of population of Oregon is located in Linn County, in the city of Lyons.[80] More than 57% of the state's population lives in the Portland metropolitan area.[81]

As of 2004, Oregon's population included 309,700 foreign-born residents (accounting for 8.7% of the state population).

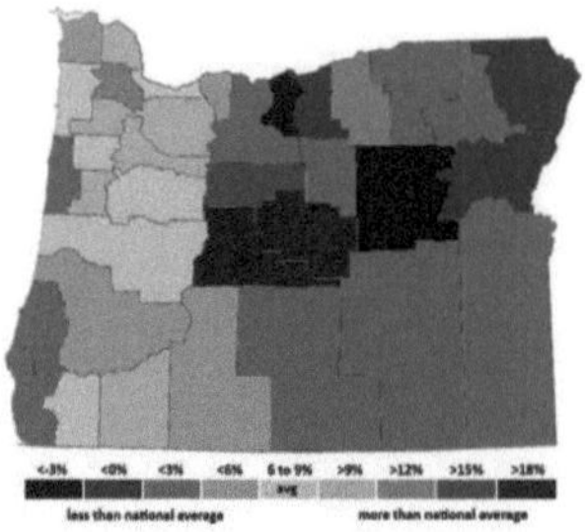

Population Growth by County, 2000–2007. Green counties grew faster than the national average, while purple counties grew more slowly or, in a few cases, lost population.

Demographics of Oregon (csv) [82]

By race	White	Black	AIAN*	Asian	NHPI*
2000 (total population)	93.45%	2.17%	2.54%	3.75%	0.48%
2000 (Hispanic only)	7.63%	0.17%	0.32%	0.10%	0.05%
2005 (total population)	92.95%	2.38%	2.44%	4.25%	0.50%
2005 (Hispanic only)	9.38%	0.24%	0.34%	0.11%	0.05%
Growth 2000–05 (total population)	5.85%	16.64%	2.45%	20.78%	10.87%
Growth 2000–05 (non-Hispanic only)	3.63%	13.63%	0.62%	20.75%	10.26%
Growth 2000–05 (Hispanic only)	30.84%	52.63%	15.25%	21.84%	16.42%
* AIAN is American Indian or Alaskan Native; NHPI is Native Hawaiian or Pacific Islander					

The largest ancestry groups in the state are:[83]

- 22.5% German
- 14.0% English
- 13.2% Irish
- 8.4% Scandinavian: (1.2% Danish, 3.1% Swedish, & 4.1% Norwegian American)
- 5.0% American
- 3.9% French
- 3.7% Italian
- 3.6% Scottish
- 2.7% Scots-Irish
- 2.6% Dutch
- 1.9% Polish
- 1.4% Russian
- 1.1% Welsh

The largest reported ancestry groups in Oregon are: German (22.5%), English (14.0%), Irish (13.2%), Scandinavian (8.4%) and American (5.0%). Approximately 62% of Oregon residents are wholly or partly of English, Welsh, Irish or Scottish ancestry. Most Oregon counties are inhabited principally by residents of Northwestern-European ancestry. Concentrations of Mexican-Americans are highest in Malheur and Jefferson counties.

The majority of the diversity in Oregon is in the Portland metropolitan area.

Future projections

Projections from the U.S. Census Bureau show Oregon's population increasing to 4,833,918 by 2030, an increase of 41.3% compared to the state's population of 3,421,399 in 2000.[84] The state's own projections forecast a total population of 5,425,408 in 2040.[85]

Religious and secular communities

Major religious affiliations of the people of Oregon are:[86]

- Christian — 67%
  - Protestant — 47%
    - Evangelical — 30%
    - Mainline— 16%
    - Other Protestant — 1%
  - Roman Catholic — 14%
  - Latter Day Saint / Mormon — 5%
  - Other Christian traditions — 1%
- Unaffiliated — 27%
- Buddhist — 2%
- Jewish — 1%
- Muslim — 0.5%
- Other Religions — 2%

The largest denominations by number of adherents in 2000 were the Roman Catholic Church with 348,239; the Church of Jesus Christ of Latter-day Saints with 104,312 (144,808 year-end 2007); and the Assemblies of God with 49,357.[87]

In a 2009 Gallup poll, 69% of Oregonians identified themselves as being Christian.[88] Most of the remainder of the population had no religious affiliation; the 2008 American Religious Identification Survey (ARIS) placed Oregon as tied with Nevada in fifth place of U.S. states having the highest percentage of residents identifying themselves as "non-religious", at 24 percent.[89] [90] Secular organizations include the Center for Inquiry (CFI), the Humanists of Greater Portland (HGP), and the United States Atheists (USA).

During much of the 1990s, a group of conservative Christians formed the Oregon Citizens Alliance, and unsuccessfully tried to pass legislation to prevent "gay sensitivity training" in public schools and legal benefits for homosexual couples.[91]

Oregon also contains the largest community of Russian Old Believers to be found in the United States.[92] The Northwest Tibetan Cultural Association is headquartered in Portland, and the New Age film *What the Bleep Do We Know!?* was filmed and had its premiere in Portland. There are an estimated 6,000 to 10,000 Muslims in Oregon, most of whom live in and around Portland.[93]

Education

Primary and secondary

As of 2010, the state had 561,698 students in public primary and secondary schools.[94] There were 197 public school districts at that time, served by 20 education service districts.[94] The five largest school districts as of 2007 were: Portland Public Schools (46,262 students); Salem-Keizer School District (40,106); Beaverton School District (37,821); Hillsboro School District (20,401); and Eugene School District (18,025).[95]

Colleges and universities

Public

The Memorial Union at OSU.

The Oregon University System supports seven public universities and one affiliate in the state. The University of Oregon in Eugene is Oregon's flagship liberal arts institution,[96] and was the state's only nationally ranked university by *U.S. News & World Report*.[97] Oregon State University is located in Corvallis and holds the distinction of being the state's flagship research university with top ranked programs in science, engineering, and agriculture. The university is also the state's highest ranking university/college in a world survey of academic merit.[98]

The state's urban Portland State University has Oregon's largest enrollment. The state has three regional universities: Western Oregon University in Monmouth, Southern Oregon University in Ashland, and Eastern Oregon University in La Grande. The Oregon Institute of Technology has its campus in Klamath Falls. The quasi-public Oregon Health & Science University (OHSU) includes medical, dental, and nursing schools, and graduate programs in biomedical sciences in Portland and a science and engineering school in Hillsboro. It rated 2nd among US best medical schools for primary care based on research by The Med School 100.[99]

Oregon has historically struggled to fund higher education. Recently, Oregon has cut its higher education budget over 2002–2006 and now Oregon ranks 46th in the country in state spending per student. However, 2007 legislation forced tuition increases to cap at 3% per year, and funded the OUS far beyond the requested governor's budget.[100]

The state also supports 17 community colleges.

Private

Oregon is home to a wide variety of private colleges. The University of Portland and Marylhurst University are Catholic institutions in the Portland area. Reed College; Concordia University; Lewis & Clark College; Multnomah Bible College; Portland Bible College; Warner Pacific College; Cascade College; the National College of Natural Medicine; and Western Seminary, a theological graduate school; are also in Portland. Pacific University is in the Portland suburb of Forest Grove.

There are also private colleges farther south in the Willamette Valley. McMinnville has Linfield College, while nearby Newberg is home to George Fox University. Salem is home to two private schools: Willamette University (the state's oldest, established during the provisional period) and Corban University. Also located near Salem is Mount Angel Seminary, one of America's largest Roman Catholic seminaries. Eugene is home to three private colleges: Northwest Christian University, New Hope Christian College, and Gutenberg College.

Sports

Oregon is home to two professional sports teams which are both based in Portland: Portland Trail Blazers of the NBA and the Portland Timbers of MLS.[101]

The Rose Garden during a Portland Trail Blazers game

Until 2011, the only major professional sports team in Oregon was the Portland Trail Blazers of the National Basketball Association. From the 1970s to the 1990s, the Blazers were one of the most successful teams in the NBA in terms of both win-loss record and attendance. In the early 21st century, the team's popularity declined due to personnel and financial issues, but revived after the departure of controversial players and the acquisition of new players such as Brandon Roy, LaMarcus Aldridge, and Greg Oden.[102] [103]

The Blazers play in the Rose Garden in Portland's Lloyd District, which also is home to the Portland Winterhawks of the junior-league Western Hockey League.[104]

The Timbers play at Jeld-Wen Field, which is just west of downtown Portland. The Timbers repurposed the formerly multi-use stadium into a soccer-only configuration in fall 2010, increasing the seating in the process.[105]

Portland has had minor-league baseball teams in the past, including the Portland Beavers and Portland Rockies, who played most recently at PGE Park. Portland has also actively pursued a Major League Baseball team.[106] It was announced in March 2009 that the Portland Timbers will begin MLS play starting in 2011.[101]

Eugene and Salem also have minor-league baseball teams. The Eugene Emeralds and the Salem-Keizer Volcanoes both play in the Single-A Northwest League.[107] Oregon also has four teams in the fledgling International Basketball League: the Portland Chinooks, Central Oregon Hotshots, Salem Stampede, and the Eugene Chargers.[108]

The Oregon State Beavers and the University of Oregon Ducks football teams of the Pacific-12 Conference meet annually in the Civil War. Both schools have had recent success in other sports as well: Oregon State won back-to-back college baseball championships in 2006 and 2007,[109] and the University of Oregon won back-to-back NCAA men's cross country championships in 2007 and 2008.[110]

Sister states

- People's Republic of China, Fujian Province – 1984[111]
- Republic of China, Taiwan – 1985[111]
- Japan, Toyama Prefecture – 1991[111] [112]
- Republic of Korea, Jeollanam-do Province – 1996[111] [112]
- Iraq, Iraqi Kurdistan – 2005[113]

See also

- Outline of Oregon
- Index of Oregon-related articles

References

[1] Calvin Hall (January 30, 2007). "English as Oregon's official language? It could happen" (http://dailyemerald.com/2007/01/30/english-as-oregons-official-language-it-could-happen/). *The Oregon Daily Emerald.* . Retrieved May 8, 2007.

[2] "Annual Estimates of the Population for the United States, Regions, States, and Puerto Rico: April 1, 2010 to July 1, 2012" (http://www.census.gov/popest/data/national/totals/2012/index.html) (CSV). *2012 Population Estimates*. United States Census Bureau, Population Division. December 2012. . Retrieved December 24, 2012.

[3] "Mount Hood Highest Point" (http://www.ngs.noaa.gov/cgi-bin/ds_mark.prl?PidBox=RC2244). *NGS data sheet.* U.S. National Geodetic Survey. . Retrieved October 24, 2011.

[4] "Elevations and Distances in the United States" (http://egsc.usgs.gov/isb/pubs/booklets/elvadist/elvadist.html). United States Geological Survey. 2001. . Retrieved October 24, 2011.

[5] Elevation adjusted to North American Vertical Datum of 1988.

[6] Map of Oregon. (http://nationalatlas.gov/printable/images/pdf/reference/pagegen_or.pdf) Reston, Virginia: Interior Geological Survey, 2004

[7] Oregon Fast Facts. (http://traveloregon.mediaroom.com/index.php?s=23&item=33) Travel Oregon.

[8] Oregon Almanac (http://www.bluebook.state.or.us/facts/almanac/almanac04.htm)

[9] Where does the name "Oregon" come from? (http://bluebook.state.or.us/misc/about/faq.htm#oregon) from the online edition of the Oregon Blue Book

[10] Elliott, T.C. (June 1921). "The Origin of the Name Oregon" (http://books.google.com/books?id=P-oXAAAAYAAJ&pg=PA99). *Oregon Historical Quarterly* **XXIII** (2): 99–100. ISSN 0030-4727. OCLC 1714620. . Retrieved October 11, 2010.

[11] Miller, Joaquin (1904). "Sea of Silence", *Sunset*, 396(13):5. (http://books.google.com/books?id=GWhpzV9XeQUC&dq=sea of silence miller&pg=PA395#v=onepage&q=sea of silence miller&f=false)

[12] "Oregon" (http://www.merriam-webster.com/dictionary/oregon). Merriam-Webster Online Dictionary. . Retrieved September 14, 2006.

[13] Banks, Don (April 21, 2002). "Harrington confident about Detroit QB challenge." (http://sportsillustrated.cnn.com/inside_game/don_banks/news/2002/04/21/harrington_intro/) *Sports Illustrated.*

[14] Bellamy, Ron (October 6, 2003). "See no evil, hear no evil" (http://news.google.com/newspapers?id=MFFWAAAAIBAJ&sjid=qusDAAAAIBAJ&dq=Joey Harrington scoffs at criticism as he struggles to right the Lions&pg=3329,1524986). *The Register-Guard.* . Retrieved June 1, 2011.

[15] "Yellow/Green ORYGUN Block Letter Outside Decal" (http://spiritduck.uoduckstore.com/Yellow_Green_ORYGUN_Block_Letter_Outside_Decal_p/76386407024.htm). UO Duck Store. . Retrieved August 3, 2011.

[16] Robbins, William G. (2005). *Oregon: This Storied Land.* Oregon Historical Society Press. ISBN 0-87595-286-0.

[17] "Oregon History: Great Basin" (http://bluebook.state.or.us/cultural/history/history04.htm). *Oregon Blue Book.* Oregon State Archives. . Retrieved September 2, 2007.

[18] "Oregon History: Northwest Coast" (http://bluebook.state.or.us/cultural/history/history02.htm). *Oregon Blue Book.* Oregon State Archives. . Retrieved September 2, 2007.

[19] "Confederated Tribes of the Grand Ronde: Culture" (http://www.grandronde.org/culture/). . Retrieved September 2, 2007.

[20] "Oregon History: Columbia Plateau" (http://bluebook.state.or.us/cultural/history/history03.htm). *Oregon Blue Book.* Oregon State Archives. . Retrieved September 2, 2007.

[21] "Chronology of events, 1543–1859" (http://arcweb.sos.state.or.us/echoes/chronology.html). *Echoes of Oregon History Learning Guide.* Oregon State Archives. . Retrieved August 11, 2011.

[22] Loy, Willam G.; Stuart Allan, Aileen R. Buckley, James E. Meacham (2001). *Atlas of Oregon.* University of Oregon Press. pp. 12–13. ISBN 0-87114-101-9.

[23] McLagan, Elizabeth (1980). *A Peculiar Paradise*. Georgian Press. ISBN 0-9603408-2-3.

[24] United States—States; and Puerto Rico: GCT-PH1-R. Population, Housing Units, Area, and Density. (http://factfinder.census.gov/servlet/GCTTable?_bm=n&_lang=en&mt_name=DEC_2000_SF1_U_GCTPH1R_US9S&format=US-9S&_box_head_nbr=GCT-PH1-R&ds_name=DEC_2000_SF1_U&geo_id=01000US) U.S. Census Bureau. Retrieved on March 28, 2008.

[25] "Elevations and Distances in the United States" (http://egsc.usgs.gov/isb/pubs/booklets/elvadist/elvadist.html). U.S Geological Survey. April 29, 2005. . Retrieved November 7, 2006.

[26] "Crater Lake National Park" (http://www.nps.gov/crla/index.htm). U.S. National Park Service. . Retrieved November 22, 2006.

[27] "D River State Recreation Site" (http://www.oregonstateparks.org/park_214.php). *Oregon Parks and Recreation Department.* . Retrieved May 11, 2007.

[28] "World's Shortest River" (http://montanakids.com/db_engine/presentations/presentation.asp?pid=192). *Travel Montana.* . Retrieved May 11, 2007.

[29] "Mill Ends Park" (http://www.portlandonline.com/parks/finder/index.cfm?PropertyID=265&action=ViewPark). *Portland Parks and Recreation.* . Retrieved May 11, 2007.
[30] Beale, Bob (April 10, 2003). "Humungous fungus: world's largest organism?" (http://www.abc.net.au/science/news/enviro/EnviroRepublish_828525.htm) Environment & Nature News, ABC Online. Accessed January 2, 2007.
[31] Western States Data Public Land Acreage (http://www.wildlandfire.com/docs/2007/western-states-data-public-land.htm) (November 13, 2007).
[32] "2010 Census Redistricting Data" (http://factfinder2.census.gov/faces/tableservices/jsf/pages/productview.xhtml?pid=DEC_10_PL_GCTPL2.ST13&prodType=table). US Census Bureau. . Retrieved March 15, 2011.
[33] 50 Fastest-Growing Metro Areas Concentrated in West and South. (http://www.census.gov/newsroom/releases/archives/population/cb07-51.html) U.S. Census Bureau 2005. Retrieved October 16, 2007.
[34] Allen, Cain (2006). "The Oregon History Project- A Pacific Republic" (http://www.ohs.org/education/oregonhistory/historical_records/dspDocument.cfm?doc_ID=92F96702-9FCF-BAC0-3411BF129F1F23DA). Oregon Historical Society. Retrieved October 8, 2010.
[35] Oregon Secretary of State. "A Brief History of the Oregon Territorial Period" (http://arcweb.sos.state.or.us/echoes/history.html). State of Oregon. . Retrieved August 9, 2006.
[36] "Constitution of Oregon (Article V)" (http://bluebook.state.or.us/state/constitution/constitution05.htm). *Oregon Blue Book.* State of Oregon. 2007. . Retrieved March 12, 2008.
[37] ORS 653.025 (http://www.leg.state.or.us/ors/653.html).
[38] "November 2, 2004, General Election Abstract of Votes: STATE MEASURE NO. 36" (http://www.sos.state.or.us/elections/nov22004/abstract/m36.pdf). Oregon Secretary of State. . Retrieved November 17, 2008.
[39] Bradbury, Bill (November 6, 2007). "Official Results – November 6, 2007 Special Election" (http://www.sos.state.or.us/elections/nov62007/). *Elections Division.* Oregon Secretary of State. . Retrieved December 27, 2008.
[40] "November 7, 2006, general election abstracts of votes: state measure no. 39" (http://www.sos.state.or.us/elections/nov72006/results/m39.pdf). State of Oregon. . Retrieved March 12, 2011.
[41] S. Spacek, 2011 American State Litter Scorecard: New Rankings for An Increasingly Environmentally-Concerned Populace.
[42] "United States Bankruptcy Court, District of Oregon" (http://www.orb.uscourts.gov/). U.S. Courts. . Retrieved December 14, 2008.
[43] Leip, David. "2008 presidential general election results" (http://uselectionatlas.org/RESULTS/). Retrieved October 12, 2010.
[44] Silver, Nate (May 17, 2008). "Oregon: Swing state or latte-drinking, Prius-driving lesbian commune?" (http://www.fivethirtyeight.com/2008/05/oregon-swing-state-or-latte-drinking.html). FiveThirtyEight.com. .
[45] "State Initiative and Referendum Summary" (http://www.iandrinstitute.org/statewide_i&r.htm). State Initiative & Referendum Institute at USC. . Retrieved November 27, 2006.
[46] "Eighth Annual Report on Oregon's Death with Dignity Act" (http://www.oregon.gov/DHS/ph/pas/docs/year8.pdf) (PDF). Oregon Department of Human Services. March 9, 2006. . Retrieved June 11, 2007.
[47] "Voting In Oregon – Vote By Mail." (http://www.co.multnomah.or.us/dbcs/elections/election_information/voting_in_oregon.shtml) Multnomah County, Oregon.
[48] Comparison of State and Local Government Revenue and Debt in the United States (http://www.usgovernmentrevenue.com/state_rev_summary.php?chart=Z0&year=2010&units=d&rank=a) Bureau of Economic Analysis.
[49] McNab, W. Henry; Avers, Peter E (July 1994). *Ecological Subregions of the United States.* Chapter 24. (http://www.fs.fed.us/land/pubs/ecoregions/) U.S. Forest Service and Dept. of Agriculture.
[50] "Industry Facts" (http://oregonwine.org/press/StateWineFacts2005.pdf) (PDF). Oregon Winegrowers Association. . Retrieved November 23, 2006.
[51] "Oregon Forest Facts: 25-Year Harvest History" (http://www.oregonforests.org/factbook/Harvest_History(24).html). Oregon Forest Resources Institute. . Retrieved March 7, 2007.
[52] "Forest Economics and Employment" (http://www.oregonforests.org/factbook/economics(29_30).html). Oregon Forest Resources Institute. . Retrieved March 8, 2007.
[53] "Oregon economy" (http://www.e-referencedesk.com/resources/state-economy/oregon.html). e-ReferenceDesk. Retrieved November 5, 2010.
[54] "Oregon's Beer Week gets under way." (http://goliath.ecnext.com/coms2/gi_0198-242714/Oregon-s-Beer-Week-gets.html). Knight-Ridder Tribune News Service. July 5, 2005. . Retrieved October 22, 2007.
[55] Don Hamilton (July 19, 2002). "Matt Groening's Portland" (http://portlandtribune.com/news/story.php?story_id=12392). *The Portland Tribune.* . Retrieved March 7, 2007.
[56] Rogoway, Mike (January 15, 2009). "Intel profits slide, company uncertain about outlook" (http://www.oregonlive.com/business/index.ssf/2009/01/intel_profits_slide_company_un.html). *The Oregonian.* . Retrieved January 16, 2009.
[57] "Genentech Selects Hillsboro" (http://www.hillchamber.org/memberservices/in_the_news.asp#Genentech). Hillsboro Chamber of Commerce. . Retrieved March 21, 2007.
[58] "Bright spots amid the turmoil" (http://www.oregonlive.com/business/oregonian/index.ssf?/base/business/1199161505105830.xml&coll=7). *The Oregonian*: p. D3. January 1, 2008. . Retrieved January 1, 2007.
[59] http://www.google.com/publicdata/explore?ds=z1ebjpgk2654c1_&met_y=unemployment_rate&idim=state:ST410000&fdim_y=seasonality:S&dl=en&hl=en&q=unemployment+rate+oregon

[60] Rogoway, Mike (January 12, 2009). "Oregon's largest private employer" (http://blog.oregonlive.com/siliconforest/2009/01/oregons_largest_private_employ.html). The Oregonian. . Retrieved January 25, 2011.
[61] "Government Finance: State Government" (http://bluebook.state.or.us/state/govtfinance/govtfinance01.htm). *Oregon Blue Book.* . Retrieved June 20, 2007.
[62] Har, Janie (June 20, 2007). "Your loss is state's record game" (http://0-docs.newsbank.com.catalog.multcolib.org/openurl?ctx_ver=z39.88-2004&rft_id=info:sid/iw.newsbank.com:NewsBank:ORGB&rft_val_format=info:ofi/fmt:kev:mtx:ctx&rft_dat=119E94A840D2FBB8&svc_dat=InfoWeb:aggregated4&req_dat=0D10F2CADB4B24C0). *The Oregonian.* . Retrieved June 20, 2007.
[63] "State Sales Tax Rates" (http://www.taxadmin.org/FTA/rate/sales.html). Federation of Tax Administrators. January 1, 2008. . Retrieved April 2, 2008.
[64] "25th Anniversary Issue: 1993" (http://www.wweek.com/html/25-1993.html). *Willamette Week.* . Retrieved June 11, 2007.
[65] "Initiative, Referendum and Recall: 1988–1995" (http://bluebook.state.or.us/state/elections/elections21.htm). *Oregon Blue Book.* State of Oregon. . Retrieved June 11, 2007.
[66] Sheketoff, Charles (March 27, 2007). "As Maryland Goes, So Should Oregon" (http://salem-news.com/articles/march272007/oregon_mrlnd_32707.php). *Salem News.* . Retrieved June 10, 2007.
[67] "Oregon ranks 41st in taxes per capita" (http://www.bizjournals.com/portland/stories/2006/03/27/daily28.html). Portland Business Journal. March 31, 2006. . Retrieved June 10, 2007.
[68] "Food and Beverage Tax" (http://www.ashland.or.us/Page.asp?NavID=9180). City of Ashland. . Retrieved June 10, 2007.
[69] "Oregon's 2% Kicker" (http://www.leg.state.or.us/comm/lro/rr02-07.pdf) (PDF). *Oregon State Leglislative Review Office.* . Retrieved June 10, 2007.
[70] Cain, Brad (March 2, 2006). "Kicker tax rebate eyed to help school and state budgets" (http://www.katu.com/news/3617476.html). KATU.com. . Retrieved June 10, 2006.
[71] "2 Percent Surplus Refund (Kicker) History" (http://www.oregon.gov/DOR/NEWS/docs/kicker.pdf) (PDF). State of Oregon. . Retrieved June 10, 2007.
[72] Cooper, Matt (March 9, 2007). "County may scrub income tax" (http://www2.registerguard.com/cms/index.php/static/search/archive/?q=County+may+scrub+income+tax). *The Register-Guard.* . Retrieved March 9, 2007.
[73] "2006 Oregon full-year resident tax form instructions." (http://www.oregon.gov/DOR/PERTAX/docs/2006Forms/101-043-06.pdf) Oregon.Gov.
[74] "Oregon" (http://www.census.gov/dmd/www/resapport/states/oregon.pdf). *Resident Population and Apportionment of the U.S. House of Representatives.* U.S. Census Bureau. December 27, 2000. . Retrieved August 28, 2009.
[75] "Annual Population Estimates" (http://www.pdx.edu/prc/annualorpopulation.html). Portland State University Population Research Center. . Retrieved March 3, 2008.
[76] "2010 census profiles: Oregon and its counties" (http://www.pdx.edu/sites/www.pdx.edu.prc/files/media_assets/2010_PL94_counties_updated.pdf). Portland State University Population Research Center. . Retrieved May 15, 2011.
[77] "Oregon QuickFacts" (http://quickfacts.census.gov/qfd/states/41000.html). U.S. Census Bureau. .
[78] "Oregon - Race and Hispanic Origin: 1850 to 1990" (http://www.census.gov/population/www/documentation/twps0056/twps0056.html). U.S. Census Bureau. .
[79] " Americans under age 1 now mostly minorities, but not in Ohio: Statistical Snapshot (http://www.cleveland.com/datacentral/index.ssf/2012/06/americas_under_age_1_populatio.html)". *The Plain Dealer.* June 3, 2012.
[80] "Population and Population Centers by State: 2000" (http://www.census.gov/geo/www/cenpop/statecenters.txt). U.S. Census Bureau. . Retrieved November 23, 2006.
[81] "JULY 1, 2006 Population estimates for Metropolitan Combined Statistical Areas" (http://web.archive.org/web/20070920212740/http://www.census.gov/population/www/estimates/metro_general/2006/CBSA-EST2006-01.csv) (csv). U.S. Census Bureau. Archived from the original (http://www.census.gov/population/www/estimates/metro_general/2006/CBSA-EST2006-01.csv) on September 20, 2007. . Retrieved October 19, 2007.
[82] http://www.census.gov/popest/states/asrh/tables/SC-EST2005-03-41.csv
[83] "2006–2008 American Community Survey 3-year estimates." (http://factfinder.census.gov/servlet/ADPTable?_bm=y&-context=adp&-qr_name=ACS_2008_3YR_G00_DP3YR2&-ds_name=ACS_2008_3YR_G00_&-tree_id=3308&-redoLog=false&-_caller=geoselect&-geo_id=04000US41&-format=&-_lang=en) U.S. Census Bureau.
[84] United States Census Bureau (April 21, 2005). "Interim Projections of the Total Population for the United States and States: April 1, 2000 to July 1, 2030" (http://www.census.gov/population/projections/SummaryTabA1.pdf). . Retrieved August 18, 2010.
[85] Office of Economic Analysis (April 2004). "State and County Population Forecasts and Components of Change, 2000 to 2040" (http://www.oregon.gov/DAS/OEA/docs/demographic/pop_components.xls). Oregon Department of Administrative Services. . Retrieved August 25, 2010.
[86] "U.S. Religious Landscape Survey." (http://religions.pewforum.org/maps) The Pew Forum on Religion & Public Life. Retrieved February 5, 2010.
[87] "State membership report- Oregon." (http://www.thearda.com/mapsReports/reports/state/41_2000.asp) The Association of Religion Data Archives.

[88] Newport, Frank (August 7, 2009). "Religious identity: States differ widely." (http://www.gallup.com/poll/122075/Religious-Identity-States-Differ-Widely.aspx#2) Gallup. Retrieved December 23, 2009.
[89] Kosmin, Barry A.; Keysar, Ariela. "American Religious Identification Survey." (http://www.americanreligionsurvey-aris.org/reports/ARIS_Report_2008.pdf) Hartford: Trinity College. Retrieved December 23, 2009.
[90] Kosmin, Barry A.; Keysar, Ariela with Cragun, Ryan; Navarro-Rivera, Juhem. "American nones: The profile of the no religion population." (http://www.americanreligionsurvey-aris.org/reports/NONES_08.pdf) Hartford: Trinity College. Retrieved December 23, 2009.
[91] Wentz, Patty (February 11, 1998). "He's back." (http://wweek.com/html/cover021198.html) *Willamette Week*. Retrieved on March 14, 2008.
[92] Binus, Joshua. "The Oregon History Project: Russian Old Believers." (http://www.ohs.org/education/oregonhistory/historical_records/dspDocument.cfm?doc_ID=764E6BED-FFC4-C034-9A5563F41CE37080) Oregon Historical Society. Retrieved on March 14, 2008.
[93] Islam in Oregon and America—The Facts (http://web.archive.org/web/20031029021946/http://www.metpdx.org/resources/)
[94] "Oregon Blue Book: Oregon Almanac: Native Americans to shoes, oldest." (http://bluebook.state.or.us/facts/almanac/almanac05.htm) Oregon Secretary of State. Retrieved on January 5, 2012.
[95] "Oregon public school enrollment increases during 2007–08." (http://www.ode.state.or.us/news/releases/?yr=0000&kw=&rid=610) Oregon Department of Education. Retrieved on March 28, 2008.
[96] Wood, Shelby Oppel (May 1, 2006). "UO weighs new diversity plan amid simmering racial tensions" (http://www.oregonlive.com/search/oregonian/). *The Oregonian*. .
[97] USNews.com: America's Best Colleges 2008: National Universities: Best Schools (http://colleges.usnews.rankingsandreviews.com/usnews/edu/college/rankings/brief/t1natudoc_brief.php)
[98] "Top 500 World Universities" (http://www.arwu.org/). . Retrieved October 3, 2012.
[99] Good University Ranking Guide (http://whichuniversitybest.blogspot.com/2010/03/ohsu-school-of-medicine.html)
[100] "Higher education get higher priority." (http://media.www.dailyemerald.com/media/storage/paper859/news/2007/06/29/News/Higher.Education.Gets.Higher.Priority-2919794.shtml) *The Oregon Daily Emerald*, June 29, 2007. Retrieved July 8, 2007.
[101] "MLS awards team to Portland for 2011." (http://portlandtimbers.com/newsroom/headlines/index.html?article_id=1108) Portland Timbers, March 20, 2009.
[102] Smith, Sam (October 18, 2006). "Blazers stalled until bad apples go" (http://www.msnbc.msn.com/id/15321476/). MSNBC. . Retrieved January 15, 2008.
[103] Mejia, Tony (October 13, 2007). "Oden's loss hurts, but team in good hands" (http://www.cbsnews.com/stories/sportsline/main10406427.shtml). CBSNews.com. . Retrieved January 15, 2008.
[104] "Rose Quarter Venues" (http://www.rosequarter.com/RoseQuarter/Venues/tabid/84/Default.aspx). RoseQuarter.com. . Retrieved January 15, 2008.
[105] "PGE Park Teams and Events" (http://www.pgepark.com/stadium/events/). PGEPark.com. . Retrieved January 15, 2008.
[106] "Oregon Stadium Campaign" (http://www.oregonstadiumcampaign.com/). Oregon Stadium Campaign. . Retrieved January 14, 2008.
[107] "Northwest League." (http://web.minorleaguebaseball.com/index.jsp?sid=l126) Minor League Baseball. Retrieved January 15, 2008.
[108] "International Basketball League." (http://www.iblhoopsonline.com/) International Basketball League. Retrieved January 15, 2008.
[109] Beseda, Jim (August 12, 2010). "Oregon State baseball: Coach Pat Casey praises ex-Beaver Darwin Barney" (http://blog.oregonlive.com/behindbeaversbeat/2010/08/oregon_state_baseball_coach_pa_2.html). *The Oregonian* (Portland, Oregon). Retrieved October 8, 2010.
[110] The Associated Press (January 8, 2009). "Oregon men, Washington women win titles" (http://sports.espn.go.com/ncaa/news/story?id=3723958). ESPN. Retrieved October 8, 2010.
[111] Van Winkle, Teresa (June 2008). "Background brief on international trade." (http://www.leg.state.or.us/comm/commsrvs/background_briefs2008/briefs/EconomyBusinessLabor/InternationalTrade.pdf) State of Oregon. Retrieved July 21, 2008.
[112] "Governor's mission to Asia will stress trade and cultural ties." (http://archivedwebsites.sos.state.or.us/Governor_Kitzhaber_2003/governor/press/p951024.htm) Oregon Secretary of State, October 24, 1995. Retrieved April 2, 2008.
[113] Senate Concurrent Resolution 3. Oregon Laws 2005 (http://www.leg.state.or.us/05orlaws/sessresmem.dir/scr0003ses.htm)

Further reading

- *Excursion to the Oregon* by John Kirk Townsend (http://www.secstate.wa.gov/history/publications_detail.aspx?p=33)
- *New map of Texas, Oregon and California with the regions adjoining, compiled from the more recent authorities* by Samuel Augustus Mitchell (http://www.secstate.wa.gov/history/maps_detail.aspx?m=14)
- *Accompaniment to Mitchell's New map of Texas, Oregon, and California, with the regions adjoining* by Samuel Augustus Mitchell (http://www.secstate.wa.gov/history/publications_detail.aspx?p=26)
- O'Hara, E. (1911). Oregon (http://www.newadvent.org/cathen/11288a.htm). In the *Catholic Encyclopedia*. New York: Robert Appleton Company. Retrieved July 25, 2009, from New Advent.

External links

- State of Oregon (http://www.oregon.gov/) (official website)
- Oregon Blue Book (http://bluebook.state.or.us/), the online version of the state's official directory and fact book
- TravelOregon.com (http://www.TravelOregon.com/) an official website of the Oregon Tourism Commission
- Oregon Historical Society (http://www.ohs.org/)
- Oregon State Databases (http://wikis.ala.org/godort/index.php/Oregon), an annotated list, in wiki form, of searchable databases produced by Oregon state agencies and compiled by the Government Documents Roundtable of the American Library Association
- Real-time, geographic, and other scientific resources of Oregon (http://www.usgs.gov/state/state.asp?State=OR) from the United States Geological Survey
- Oregon Quickfacts (http://quickfacts.census.gov/qfd/states/41000.html) from the United States Census Bureau
- Oregon State Facts (http://www.ers.usda.gov/StateFacts/OR.htm) from the United States Department of Agriculture
- Oregon (http://www.dmoz.org/Regional/North_America/United_States/Oregon/) at the Open Directory Project
- Geographic data related to Oregon (http://www.openstreetmap.org/browse/relation/165476) at OpenStreetMap

Crook_County,_Oregon

Crook County, Oregon

Crook County Courthouse in Prineville

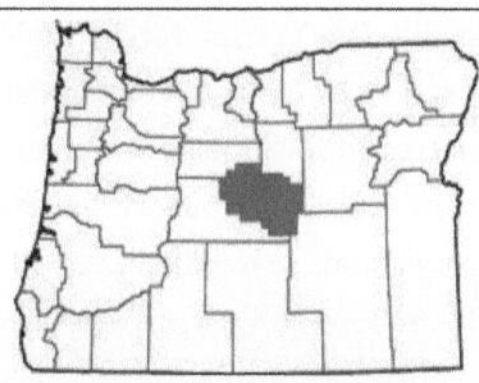

Location in the state of Oregon

Oregon's location in the U.S.

Founded	October 24, 1882
Seat	Prineville
Area **- Total** **- Land** **- Water**	2987 sq mi (7736 km²) 2979 sq mi (7716 km²) 8 sq mi (21 km²), 0.27%
Population **- (2010)** **- Density**	20978 7/sq mi (2.7/km²)
Website	www.co.crook.or.us [1]

Crook County is a county located in the U.S. state of Oregon. In 2010, its population was 20,978. It is named after George Crook, a U.S. Army officer who served in the American Civil War and various Indian Wars. The seat of the county is Prineville.

History

Access into the region at first was difficult, which discouraged settlement. The first effort to develop routes into the area was in 1862 when a supply train with cattle crossed the Scott Trail. This was also the first group of non-natives to spend the winter in central Oregon. The discovery and development of the Santiam Pass in the 1860s improved access into the area.

Crook County was formed from the southern part of Wasco County on October 24, 1882, and established Prineville as the county seat. The voters confirmed the choice of Prineville, the only incorporated town in the county, in the 1884 general election.

Logging in the Ochoco Mountains and the timber mills that accompanied it contributed to the economic and population growth of the county. The first recorded mention of a sawmill was made by George Barnes, speaking about the Swartz sawmill on Mill Creek, circa 1867.[2] [3]

Economy

Forest products, agriculture, livestock raising and recreation/tourism services constitute Crook County's total economy. Agriculture is supported by the development of irrigation districts, which permits the raising of hay, grain, mint, potatoes, and seed. Range and forest lands allow grazing for a sizable livestock industry. The Ochoco National Forest's stand of ponderosa pine is the main source of lumber. Tourism and recreation help round out the economy. Thousands of hunters, fishers, boaters, sightseers and rockhounds are annual visitors to its streams, reservoirs and the Ochoco Mountains. The Prineville Chamber of Commerce provides access to over **unknown operator: u','** acres (**unknown operator: u'strong'** km^2) of mining claims to rockhounds, who can dig for free agates, limb casts, jasper and thundereggs.

Politics

Though Crook County is the most central county in Oregon, politically it falls in line with the eastern side of the state. The majority of registered voters who are part of a political party in Crook County, as well as most counties in eastern Oregon, are members of the Republican Party.[4] In the 2008 presidential election, 61.54% of Crook County voters voted for Republican John McCain, while 35.09% voted for Democrat Barack Obama and 3.37% of voters either voted for a third-party candidate or wrote in a candidate.[5] These numbers show a small shift towards the Democratic candidate when compared to the 2004 presidential election, in which 68% of Crook Country voters voted for George W. Bush, while 30.1% voted for John Kerry, and 1.9% of voters either voted for a third-party candidate or wrote in a candidate.[6]

Crook county was formerly a Presidential bellwether county, voting with the winner since 1884, in 27 Presidential elections.[7] However, the county lost its bellwether status after voting for George H. W. Bush in 1992.[8] It further voted for the losing Presidential candidate in 1996 and 2008.

[4]

Geography

The county is located in the geographic center of Oregon. It has a total area of 2987 square miles (7736.3 km^2), of which 2979 square miles (7715.6 km^2) is land and 8 square miles (20.7 km^2) (0.27%) is water. The largest body of water in Crook County is the Prineville Reservoir. The county has been reduced from its original size of 8600 square miles (km^2) by the creation of Jefferson County in 1914 and Deschutes County in 1916. The present boundaries were established in 1927.

The oldest geological formation in Oregon is in the southeastern corner of Crook County, near its boundary with Grant County. This formation is an outcropping of Devonian limestone created from a larger reef when most of

Oregon was covered by water.

Adjacent counties

- Deschutes County, Oregon - south, west
- Jefferson County, Oregon - north
- Wheeler County, Oregon - north
- Grant County, Oregon - east
- Harney County, Oregon - southeast

National protected area

- Ochoco National Forest (part)

Demographics

Historical populations		
Census	**Pop.**	**%±**
1890	3244	—
1900	3964	22.2%
1910	9315	135.0%
1920	3424	−63.2%
1930	3336	−2.6%
1940	5533	65.9%
1950	8991	62.5%
1960	9430	4.9%
1970	9985	5.9%
1980	13091	31.1%
1990	14111	7.8%
2000	19182	35.9%
2010	20978	9.4%
[9] [10] [11]		

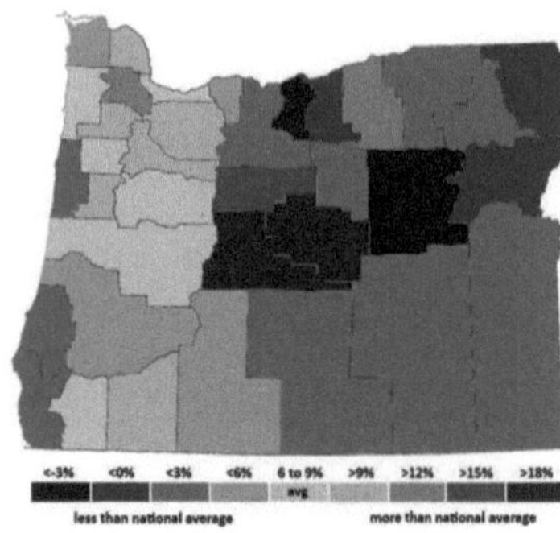

From 2000 to 2007, Crook County's population grew by 34.9%, more than three times the state average. It was the second fastest growing county in the state, after neighboring Deschutes County.

As of the census[12] of 2000, there were 19,182 people, 7,354 households, and 5,427 families residing in the county. The population density was 6 people per square mile (2/km²). There were 8,264 housing units at an average density of 3 per square mile (1/km²). The racial makeup of the county was 92.95% White, 0.04% Black or African American, 1.30% Native American, 0.43% Asian, 0.03% Pacific Islander, 3.81% from other races, and 1.43% from two or more races. 5.64% of the population were Hispanic or Latino of any race. 26.2% were of American, 14.8% German, 9.7% English and 8.9% Irish ancestry.

There were 7,354 households out of which 32.30% had children under the age of 18 living with them, 61.50% were married couples living together, 8.20% had a female householder with no husband present, and 26.20% were non-families. 21.30% of all households were made up of individuals and 9.50% had someone living alone who was 65 years of age or older. The average household size was 2.57 and the average family size was 2.96.

In the county, the population was spread out with 26.60% under the age of 18, 7.50% from 18 to 24, 25.50% from 25 to 44, 25.70% from 45 to 64, and 14.70% who were 65 years of age or older. The median age was 39 years. For every 100 females there were 99.40 males. For every 100 females age 18 and over, there were 97.30 males.

The median income for a household in the county was $35,186, and the median income for a family was $40,746. Males had a median income of $32,166 versus $22,580 for females. The per capita income for the county was $16,899. About 8.10% of families and 11.30% of the population were below the poverty line, including 13.90% of those under age 18 and 8.10% of those age 65 or over.

Communities

Incorporated city

- Prineville

Unincorporated communities

- Forest Crossing
- Lone Pine
- O'Neil
- Paulina
- Post
- Powell Butte
- Roberts
- Suplee

See also

- National Register of Historic Places listings in Crook County, Oregon

References

[1] http://www.co.crook.or.us

[2] Shaver, F. A., Arthur P. Rose, R. F. Steele, and A. E. Adams, compilers. *An Illustrated History of Central Oregon: Embracing Wasco, Sherman, Gilliam, Wheeler, Crook, Lake, & Klamath Counties*. Spokane, WA: Western Historical Publishing Co., 1905.

[3] "Ponderosa Pine Capital of the World" (http://ponderosapinecapital.org). digital collection, A. R. Bowman Memorial Museum. .

[4] State.or.us (http://www.sos.state.or.us/elections/votreg/mar09.pdf) Retrieved on 4/20/09

[5] Crook.or.us (http://co.crook.or.us/Departments/CountyClerk/2008GeneralElectionResults/tabid/1124/Default.aspx) retrieved 4/20/09

[6] City-data.com (http://www.city-data.com/county/Crook_County-OR.html) Retrieved on 4/21/09

[7] Egan, Timothy (October 13, 1992). "THE 1992 CAMPAIGN: The Bellwether County; Bellwether County Gravitates To Clinton, if Only by Default" (http://www.nytimes.com/1992/10/13/us/1992-campaign-bellwether-county-bellwether-county-gravitates-clinton-if-only.html?pagewanted=all&src=pm). New York Times. . Retrieved June 24, 2012.

[8] Fredrickson, Keith (November 4, 1992). "No Bellwether Blues in Crook County" (http://news.google.com/newspapers?nid=1243&dat=19921104&id=1RcpAAAAIBAJ&sjid=mYYDAAAAIBAJ&pg=6970,1090158). The Bend Bulletin. . Retrieved June 24, 2012.

[9] http://www.census.gov/population/www/censusdata/cencounts/files/or190090.txt

[10] http://factfinder2.census.gov

[11] http://mapserver.lib.virginia.edu/

[12] "American FactFinder" (http://factfinder.census.gov). United States Census Bureau. . Retrieved 2008-01-31.

Article Sources and Contributors

Peter_French *Source*: http://en.wikipedia.org/w/index.php?title=Peter_French *Contributors*: Aboutmovies, Akerans, Alansohn, Biglovinb, BobbieCharlton, Bobblewik, Colonies Chris, EncMstr, Finetooth, Good Olfactory, Hmains, Jsayre64, MarchOrDie, Metta Bubble, Netoholic, OfficerCampbell, Orygun, Ryulong, SCEhardt, Tesscass, UkPaolo, Valfontis, 9 anonymous edits

Pete_French_Round_Barn *Source*: http://en.wikipedia.org/w/index.php?title=Pete_French_Round_Barn *Contributors*: Aboutmovies, Appraiser, Droll, EncMstr, Hmains, Hugo999, IvoShandor, Jllm06, KudzuVine, N.M.Sheedy, NE2, Peteforsyth, Scientizzle, Valfontis, 1 anonymous edits

National_Register_of_Historic_Places *Source*: http://en.wikipedia.org/w/index.php?title=National_Register_of_Historic_Places *Contributors*: 293.xx.xxx.xx, 2over0, 72Dino, 83d40m, Abune, Aika, Aikidockd, Alan Liefting, AlbertusmagnusOP, Anarchyinyourcountry, Aoi, ArielGold, Aude, Auric, Aussieintn, Avnative, Avriette, BD2412, Bazonka, Bbedn, Bedford, Beginning, Beland, Billy Hathorn, Blainster, BlankVerse, Bobbyhasnoblueshoes, Bollar, Brholden, Briankelly25, Broadway Center for the Performing Arts, Bschorr, Bumpusjones, CBOrgatrope, Cacophony, Caltrop, Carina22, Cburnett, Centrx, Chris Light, ChrisGualtieri, CiprianAgapi, Cmasonwhf, Corker1, D6, DMG413, Da best director, Dalejarvis, Daniel Case, DanielCD, Danny, Dbastro, Dblevins2, Ddihel, Dead fishy 111, Decumanus, Doc Strange, Dogears, Doncram, Dsmdgold, Dual Freq, Ebyabe, Econrad, Ed!, Elekhh, Elkman, Eoghanacht, Erin1973, Esrever, EurekaLott, Fred Hsu, FredR, Freechild, Funandtrvl, Gadfium, Gerald Farinas, Gmatsuda, GoingBatty, Gopats92, Gregh3285, Grutness, Guliolopez, Hadlock, Hardvice, Henrygb, Hirohisat, Hmains, Holt9359, Hotcrocodile, HstNerd, Hu12, Hyuri, IbervilleMan, Infocidal, Infrogmation, Ipoellet, IvoShandor, J.H, JHunterJ, Jagvar, Jake Wartenberg, Javier martin, Jeff02, Jengod, Jesster79, Jew lover 1488, Jfknrh, Jiang, Jll, Jmundo, Jolomo, Jonathunder, Jorfer, Jpat34721, Jvhertum, KHM03, Kaibabsquirrel, Kainaw, Kaldari, Katieishot, Kbh3rd, Kenlukus, Kneiphof, Koavf, Koplimek, Kralizec!, Lars Washington, Laurascudder, LeastCommonMultiple2, Leszek Jańczuk, Leuko, LittleT889, Littlealien182, Loadmaster, Lockley, Loodog, Lotje, Lucky 6.9, Lukobe, MER-C, MJCdetroit, MONGO, MPS, Mackensen, Massalim, MattSal, Mattfox22, Mattis, Mauls, Maxatl, Meco, Michael Hardy, MichaelWheeley, Michaelenandry, MilFlyboy, Minnmass, Mjroots, Moe Epsilon, Motorrad-67, Motthoangwehuong, Mrwojo, Mulad, Mwanner, Myleslong, Natalie Erin, Natl1, Nehrams2020, Niteowlneils, Northamerica1000, Notorious4life, Ntsimp, Nunh-huh, Nyttend, OhTricky, Omicronpersei8, Onore Baka Sama, Optimist on the run, Orlady, Otrfan, Ottre, Outriggr, Parkwells, Penn Station, Petersam, Polarscribe, Postdlf, Postlebury, Preservationiscool, President Rhapsody, Quebec99, R'n'B, RTG, Rajulbat, Reywas92, Rich Farmbrough, Rjensen, Rkytop98, Robert McClenon, Rodyx, Ross Uber, Royalbroil, Rt66lt, Russianamerican1, SalomonCeb, Samoboow, Sanfranman59, Saopaulo1, Scott Burley, ScottMainwaring, ScottyBoy900Q, Sean William, Sean.t.oneill, Semiautomata, Shawn in Montreal, SidP, Skarebo, Skeezix1000, Slaterusa, Slazenger, Snigbrook, Srich32977, Station1, Stephen Burnett, Stevage, TAnthony, TVSRR, Tagishsimon, Tedder, Tedickey, Tene, The Thing That Should Not Be, Theking17825, Theopolisme, Timmer428, Timmyo101, TimothyPilgrim, Tony1, Triddle, Tryptofish, Twang, UsaAndAbuse, Useight, Valerius Tygart, Valfontis, Vaoverland, Vegaswikian, Villwock, Vrysxy, WJBscribe, Washingtonian, Wasted Time R, Wavelength, Whylimeemily, Wimstead, Wodan, Woohookitty, Yngvadottir, Yoho2001, Yoiloper, Zagalejo, ZekeMacNeil, 213 anonymous edits

Round_barn *Source*: http://en.wikipedia.org/w/index.php?title=Round_barn *Contributors*: Amity150, AnonMoos, Biruitorul, Brambleshire, Briancua, Cadastral, Chris the speller, Clariosophic, Colonies Chris, CommonsDelinker, CurlyKrakow, Doncram, Elkman, Frank Schulenburg, Gfelsman, Gilliam, Grafen, Intgr, IvoShandor, Jim Derby, Jonathunder, Juanth, Keuk, KudzuVine, Leandrod, Mild Bill Hiccup, Motorrad-67, N.M.Sheedy, Neutrality, Nyttend, Peteforsyth, Piledhigheranddeeper, Pubdog, Royalbroil, Ruhrfisch, Shenme, Spray787, Station1, Swampyank, Witan, Wulfnoth, Zoicon5, 14 anonymous edits

Burns,_Oregon *Source*: http://en.wikipedia.org/w/index.php?title=Burns%2C_Oregon *Contributors*: AXRL, Aboutmovies, Accounting4Taste, Allstarecho, Bumm13, Bunchofgrapes, Cacophony, Docu, Dravecky, ESVNSCR, EncMstr, Finetooth, Floydspinky71, Glacier109, Gtwfan52, Hike796, Hushpuckena, Jamo2008, Jllm06, Joan.ignasi.91, Johnpacklambert, Lesb246, MJCdetroit, Mais oui!, Mapcat, MathewDill, Mikevegas40, Mithridates, Mlaffs, Motthoangwehuong, NE2, Nneonneo, Nyttend, Odotptd, Ookttahbahblah, Ouzo, Owen, PDXblazers, PandiB, ParallelWolverine, Peteforsyth, Philip Trueman, PouX, Ram-Man, Rich Farmbrough, SPUI, Seglea, Tedder, Trashbag, Uyvsdi, Valfontis, Vibrantspirit, Waggers, Woohookitty, بس‌ای رس‌پ, 93 anonymous edits

Harney_County,_Oregon *Source*: http://en.wikipedia.org/w/index.php?title=Harney_County%2C_Oregon *Contributors*: 90, AXRL, Arrabbikum, AshyLarry, Backspace, Bobblewik, Breffni Whelan, Bumm13, Cacophony, CanisRufus, Carrite, Cjthellama, Curtvancil, D6, Dina, Dino, Dukeofomnium, Eaglecap Backpack, EncMstr, Finetooth, Floydspinky71, Fratrep, GrahamHardy, Ground Zero, H-stt, Hajor, Hu12, Hyperfast, JIP, Jeremy Bentham, Johnpacklambert, Jsayre64, Lesb246, Leslie Mateus, Libereco Li, Lincolnite, Llywrch, Lostwagonman, MaxwellPerkins, Michael Patrick, MojaveNC, Monegasque, Motthoangwehuong, Omnedon, Owen, Parkwells, Pearle, Peteforsyth, PurpleSunflower, Ram-Man, Rettetast, Smallbones, TUF-KAT, Tabletop, Template namespace initialisation script, Valfontis, Vis-a-visconti, Worldenc, 21 anonymous edits

Lake_County,_Oregon *Source*: http://en.wikipedia.org/w/index.php?title=Lake_County%2C_Oregon *Contributors*: 90, Aboutmovies, Ajbenj, Backspace, Bloody Brilliant, Bobblewik, Bumm13, CanisRufus, CharlieEchoTango, CommonsDelinker, D6, Davehi1, Dingar, Droll, Duff, Eaglecap Backpack, Emeraldcityserendipity, EncMstr, Finetooth, Floydspinky71, Fratrep, Fullerton, Fusillijerry86, GWBeasley, GrahamHardy, Hike796, Hmains, Hushpuckena, Hyperfast, Indefatigable, Ipoellet, JaGa, Jeff3000, Jerzy, JoeTimko, John of Reading, Jsayre64, Jyoshimi, Ken Gallager, Leslie Mateus, Libereco Li, Lightmouse, LilHelpa, Lincolnite, Llywrch, MaxwellPerkins, Monegasque, Motthoangwehuong, NE2, Nyttend, Omnedon, Orygun, Owen, PurpleSunflower, RJFJR, Ram-Man, Rich Farmbrough, Rjwilmsi, Roguefly, Rvannatta, Sgt Pinback, Shannon1, Smallbones, Sooz99, Stan Shebs, Steven Walling, T. Anthony, TUF-KAT, Tabletop, Template namespace initialisation script, Themfromspace, Valfontis, Woohookitty, Worldenc, Yrithinnd, Zeimusu, 17 anonymous edits

Deschutes_County,_Oregon *Source*: http://en.wikipedia.org/w/index.php?title=Deschutes_County%2C_Oregon *Contributors*: AXRL, Another Believer, AshyLarry, Backspace, Bentley4, Boygobig, CanisRufus, CommonsDelinker, D6, Deflective, Derek R Bullamore, Eaglecap Backpack, EncMstr, EngineerScotty, Floydspinky71, Fratrep, Funnyhat, Gadfium, Glacier109, Graham87, Ground Zero, HennessyC, Husond, Hyperfast, IBook of the Revolution, Jerryki, Jesset77, Kudzu1, Kwamikagami, Leitmotiv, Leslie Mateus, Libereco Li, Lightmouse, Lincolnite, Llywrch, Magioladitis, Michaelsbll, Monegasque, Motthoangwehuong, Niceguyedc, Northwesterner1, Numbo3, Nyttend, Occhanikov, Omnedon, Orygun, Owen, Pacificoduck, Peteforsyth, PurpleSunflower, Ram-Man, Rcsprinter123, Rich Farmbrough, Rjwilmsi, Robertjohnsonrj, Sgt Pinback, Smallbones, Stratosphere, TUF-KAT, Tabletop, Template namespace initialisation script, Tim!, Tobsters777, Valfontis, Wik, WillC, Worldenc, Worm That Turned, 65 anonymous edits

Oregon *Source*: http://en.wikipedia.org/w/index.php?title=Oregon *Contributors*: (jarbarf), -Fire-, 16@r, 3.217, 5 albert square, 7george7, 7heesub, 90, A2Kafir, A7xjackass13, ACSE, AF501, AHP, Aaron Schulz, Aaron north, AbbaIkea2010, Abby, Abce2, Aboutmovies, Acalamari, AceHarding, Adam Keller, AdamjVogt, Addshore, AdjustShift, Adjwilley, Aeusoes1, Ag1246, AgentPeppermint, Ahassan05, Ahoerstemeier, Aivazovsky, Ajbenj, Akendall, Al Silonov, Al-bayda, AlanM1, Alansohn, Alberry, Ale jrb, Alexander S., Alexindigo, AlexiusHoratius, Alexmichalskileslie, Alexwcovington, Algocu, Aliceinlampyland, Aliceiscoo, Alison, Almostmidnight, Alphachimp, Altenmann, AlysTarr, Amazonien, Amccrea, American Black bear, AmiDaniel, Amsby17, Andrewlp1991, Andy Marchbanks, Andy120290, Angleterre, Antandrus, Antipex, Antodav2007, Apple1091, Aris Katsaris, Aroldis, Art LaPella, Arthuralee, Artsojourner, Arx Fortis, Ashwinr, Astropithicus, Athelwulf, Aude, AustinRyan, Av99, Averybo, Awotter, Axoplasm, BD2412, BMRR, BSveen, Babygene52, Balcer, Ballista, BalooUrsidae, Bancroft595, Barek, Bart133, Bassbonerocks, Bastin, Bbpen, Beano, Beansbaxter, Bearclawbill, Beaver1believer, Becboobear, Beginning, Beland, Belligero, BenBaker, Bender235, Bennelliott, Bertolc, Best O Fortuna, Beveal, Beyondthislife, Bhound89, Big Smooth, Big iron, Bigcheesebebbs, Biruitorul, Bizxman, Bkonrad, Black Orchid, Blaze33541, Bluedogtn, Bmv 1978, Bobblewik, Bobomejor, Bodegaard, Bomac, Bongwarrior, Bookgrrl, Booyabazooka, Borderer, Brian0918, Brianga, Broctel, Brutaldeluxe, BryceHarrington, Buaidh, Buddha24, Burgundavia, Butts99, Butwhatdoiknow, CALR, CIreland, CJLL Wright, CO, CSL, CWenger, CWii, Cacophony, Caloy24, Calvin 1998, Can't sleep, clown will eat me, Canadian-Bacon, CanadianLinuxUser, CanadianPenguin, CanisRufus, Canseb, Cantea V, Caponer, Capricorn42, Captain panda, Capuchin the Friar, Caranha, CardinalDan, Carlsotr, Caseyo1992, Cburnett, Cchow2, CenozoicEra, Centrx, CeramicReality, Ceranthor, Chadlupkes, Chawman69, ChrisRuvolo, Chrisbrownluvr60, Chrisoule, Chrisrtait, Chrnath, Chuckgeorge60, Chuggnutt, Chzz, Ciescott, Civil Engineer III, ClairSamoht, Clarkbhm, Clipdude, Cluelessness, CommonsDelinker, Conundrum22, Conversion script, Coolcaesar, CopperSquare, Coredesat, Corey.reece, Corriebertus, Coulraphobic123, Courcelles, CowboySpartan, Craigaswartout, CrazyMan10101, Crazycomputers, Crd721, Cribbswh, Cromwellt, Cwolfsheep, Cybercobra, Cyde, Cylewithak, CylonCAG, CyranoDeWikipedia, Cyrusxneil, D6, DARTH SIDIOUS 2, DCEdwards1966, DIDouglass, DJ1AM, DRMparadigm, DVD R W, Dabomb87, Dadaist6174, Dale Arnett, Daniel Royer, DanielCD, Danny, Dark jedi requiem, Darklilac, Darth stormtrooper, Dave Cohoe, DaveGorman, Daven200520, Davepape, David Justin, David Schaich, David1217, Davidzundel, Davy hopkins@yahoo.com, Dawnseeker2000, Dcornwall, Dcvgtrd, Decltype, Decumanus, Dedicated editor, Deeet, Deflective, DeltaQuad, Demi, Den fjättrade ankan, Dendodge, DerHexer, DickClarkMises, Dino, Dirtyaffliction, Discospinster, Dmsar, Doc glasgow, DomiAllStates, Dommkopf, Doug Coldwell, Doulos Christos, Dpm12, Dr. Blofeld, Draco345, Dre503, Dual Freq, Duck4life, Ducknish, Dufekin, Duff, Dwilso, Dysepsion, E0steven, EOBeav, ERcheck, ESkog, Ed g2s, Editor at Large, Edivorce, EdwardRech, Edwardsully, Ego White Tray, Ehurtley, Ekren, El C, El Cid, Elano, Elassint, Elwood j blues, Emeraldcityserendipity, Emhoo, EmilyWolff, Emmett5, Emojones, EncMstr, EncycloPetey, Engineer Bob, EngineerScotty, Entirelybs, Eochaid, Epbr123, Epolk, Epsilon60198, Erall, ErgoSum88, Eric-Wester, Eric33sb, EricSerge, Esprqii, Eu.stefan, Evice, Evrik, Excirial, Exert, Explicit, Eyquepasa, FCSundae, FactionofReds, Faithlessthewonderboy, Falcon8765, Fang Aili, Fastily, Fastrunner61, Fbolanos, Fcb981, Filiwickers, Finder Steve, Finduilas 09, Finetooth, Firien, Fishing, Flaminsky, Flatterworld, Fondue, Foofighter20x, Freak4jc, Frietjes, Friginator, FriscoKnight, Fritz Saalfeld, Fuhghettaboutit, Funnyhat, Fupa12, Fæ, GOTU62, Gaianauta, Gaius Cornelius, Gdo01, Geologyguy, GeorgeMoney, Gettingtoit, Gfoley4, Gimboid13, Glacialfox, Glasscity09, Glendoremus, Gogo Dodo, Good Olfactory, GoodDamon, Google1230, GraemeMcRae, Grafen, Graham87, Graymornings, Greekafella, GregU, Grubbiv, Gtwkndhpqu, Guliolopez, Gurch, Gwguffey, Gwynnebaer, Hadal, Haha169, Hajatvrc, Half price, HamburgerRadio, Harmil, Haukurth, Hayaji, HeLmiT, Headwes, Heegoop, HeikoEvermann, Henryodell, Higbvuyb, Highvale, Hiimkylestevens, Hlj, Hmains, Hoefer12, HorizonOmen, Hu12, Hubertfarnsworth, Huntington, Husond, Hut 8.5, IBook of the Revolution, IRP, Ice Cold Beer, Idontgotmilk, IkonicDeath, Imawesomeandilikeit, Imma Dolphin F You, Imnaha, Imran, Indefatigable, Infoporfin, Inqvisitor, Inter16, Interiot, Intothewoods29, Iop7789, Ipawn, Ipoellet, Iridescent, IsaacTheSalsaShark, Isarra, Isoperlagravitans, Isopod, ItsZippy, Ixfd64, J'onn J'onzz, J.delanoy, JFG, JForget, JTM, JYOuyang, Ja 62, JackLumber, Jacob.jose, Jacobuskaminus, Jahiegel, Jakobees, James McNally, Jauhienij, Javert, JayHenry, Jcam, Jcworsley, Jeffersonfranklin, Jellytime86, Jengod, Jennypalazio, Jeremiestrother, Jesse Milligan, Jfire, Jfricker, Jfurr1981, Jfusion, Jgilhousen, Jgwpdx, Jhendin, Jim1138, JimIrwin, JinJian, Jlhw, Jmchuff, Jms39, Jmtn20, Jnorton7558, JoJoVader12, JoanneB, JoaoRicardo, John K, John254, JohnInDC, JohnLloydScharf, JohnOwens, Johnny Mnemonic, Jojhutton, Jomasecu, Joshbzin, Joyous!, Jrat87, Jsayre64, Jschnur, Ju66l3r, Juliancolton, Julien Deveraux, Jump222, JustBear, Justrob, KGasso,

KPH2293, Kaijan, Kaiser matias, Kaltenmeyer, Kayaker, Kbdank71, Kbreth, KeithH, Ken Gallager, Kevin B12, Kevin Myers, Kevin908, Kgrad, Kiko4564, Kilowattradio, Kimchi.sg, King prime Minister, King256, Kinotama, Kintetsubuffalo, Kkailas, Kkmd, Kkv123, Kman543210, Kollision, Kozuch, Krich, Ksnow, KuBi4K, Kuchiguchi, Kudzu1, Kukini, Kumioko (renamed), Kungfuadam, Kuribosshoe, Kuru, Kwamikagami, LGagnon, LJTPerry, Lampsalot, Larrykoen, Lateg, Latitude0116, Lean.joe, LeaveSleaves, LedgendGamer, Leszek Jańczuk, LeviathanMist, Levineps, Lfh, Liface, Lifung, LightSParker, Lightmouse, Lights, Lilac Soul, Lilmiss34, Lincpalex, LineOfBlackCars, Little Mountain 5, Llywrch, Longnskinny, Lookiemookiecookie, Lookinhere, Looxix, Lord British, Loul, LuK3, Lugia2453, Luna Santin, Luong, Luuva, Luxa, M C Y 1008, MJCdetroit, MJDTed, MK8, Macarion, Mack2, Magister Mathematicae, Makaristos, Malo, Manayse1, Manuel Trujillo Berges, Marauder09, MarcoTolo, Marek69, Mark91, MarkSweep, Martarius, Martin Kozák, Masterminded, Matt Gies, Matt.T, Matt89520, Mattmcc, Mav, Maximus Fagot, Maxtremus, Mdroses, Meadey9, Menah the Great, Merovingian, Messynessie125, Metanoid, Methychroma, Mets501, Mhking, Michael Hardy, Michael Patrick, Michaelmanninghansen, Michaelzeng7, Microtonal, Midgley, Midnight Green, Mightymights, Mike Cline, MikeTizzle777, Mikevegas40, Mild Bill Hiccup, Miles530, Mimzy1990, Minesweeper, Mkeranat, Mokgen, Molly-in-md, Monterey Bay, Mothball, Moverton, Moxy, Mr Accountable, Mr Responsible, Mr.batmanperson, MrBoo, Mrtobacco, Ms2ger, Msaroff, MuZemike, Mwanner, Mxn, N734LQ, NE2, NEICenergy, Naddy, Nagasakisullenhorde, Nagelfar, Nakon, NatureA16, NawlinWiki, NellieBly, Neofan123, Neutrality, Nevin488, NewEnglandYankee, Newport Beach, Niccus, Nicknicknuke, Nightkey, Nihiltres, Niteowlneils, Njoedits, Nkrosse, Nn123645, Nobodyiswatching, Nojam75, NorCalHistory, Northamerica1000, Northwestern guy, Northwesterner1, NorwegianBlue, Notforevil, Nufy8, Nuno Tavares, Nw adelman, O.Koslowski, Oberonfitch, Oddchump007, Offkilter, Ohconfucius, Ohnoitsjamie, Oldlaptop321, Olivier, Omega65, Omicronpersei8, Onepebble, OregonD00d, OregonDuck37, OregonRepublican, Oregongrl88, Oregonians, Orphan Wiki, Osterluzei, OverMyHead, Owen, Oxymoron83, Ozette, P. S. F. Freitas, PC1996, PDXblazers, PM800, PS2pcGAMER, Paleowiki, Pantaloons799, Parhamr, Pascal.Tesson, Pass a Method, Patrick, Paul-L, Paul2.0, Pavel Vozenilek, Pekinpekin, Pennn15, Penubag, Peteforsyth, Peter G Werner, Petercoyl, PeytonWestlake, Pfly, Pgan002, Phantomsteve, Phil Boswell, Philip Trueman, Philipscranage, Piano non troppo, Picaroon, Piledhigheranddeeper, Pilotguy, Pinethicket, Pion, Plasticup, Plowboylifestyle, Poccil, PoliceCaptain, Porce, Portlandium, Postdlf, Pourlaclasse, Praline97, Presumptive, Puget Sound, Qqqqqq, Quackslikeaduck, Quintote, QuiteUnusual, Qutezuce, Qxz, R'n'B, RG2, RJaguar3, RTC, RVJ, RVRVRVRVRVRVC, Rachelthompson44, Railer 103, Rami.J241, RandalSchwartz, Randlo97, RandomP, RattusMaximus, Rawnblade132, Realm of Shadows, RedWolf, Redtitan, Reinerd, Renato Caniatti, Retired username, Rettetast, RevTenderBranson, Revas, RexNL, Rhatsa26X, Rich Farmbrough, RichardF, Richardaedwards, Rick Block, Riverstepstonegirl, RjLesch, Rjwilmsi, Rlevse, Rmhermen, Roastytoast, Robfhu, Rocketpastsix, Romanm, Romarin, Ronhjones, Root Beers, Rootbeer, Rorschach, Rossami, Rossdegenstein, Rreagan007, Rror, Rubicon, RubixA+, Ruhrfisch, Rupertslander, Ryan5091, Ryulong, SDC, SHIMONSHA, SJP, SNIyer12, STAREYe, SURIV, Sam Hocevar, Samtheboy, SandyGeorgia, Sango123, Sannse, Saopaulo1, Sarrus, Sasata, Satesclop, SatyrTN, Scarab12, Schonchin, SchuminWeb, Scientizzle, Sciurinæ, ScottMainwaring, Scriberius, Scythian1, Seb az86556, SebastianHelm, Secret (renamed), SecurityAdvisorCode17, Seqsea, Sfmontyo, Shalom Yechiel, Shanem4311, Shanes, Shankyuu, ShelfSkewed, Shii, Shoecream, Shoeofdeath, Silverrobes, Simeon, Simon Far, SimonTrew, Singularity, Sionus, Sisselpud, Skarebo, SkerHawx, Skookum1, Sky Attacker, Slambo, Sligocki, Slj, Smith03, Snigbrook, Snowolf, Soap, Some jerk on the Internet, SomeDudeWithAUserName, Someguy1221, Sonett72, South Bay, Sowelilitokiemu, Sphincter3, SportingFlyer, Spotteddogsdotorg, Spydrlink, Staffwaterboy, Staplegunther, Starnestommy, Steve Casburn, SteveSims, Steven Walling, Steven91, Stevewonder2, Sturmdrang, Sumsum2010, Sunray, SuperHamster, Supergeekfreak, Svgalbertian, Swagadactyl, Swollib, Sword Fern, Synchronism, Szyslak, T-Bone, TDogg310, TEG24601, TJ Spyke, TMC, TUF-KAT, Ta bu shi da yu, Tabletop, Tartanwizard, TartarSauce, Tbhotch, Tedder, Tehbroseff, Template namespace initialisation script, Tesscass, Thatguyflint, The Epopt, The High Fin Sperm Whale, The Obento Musubi, The Thing That Should Not Be, The Transhumanist, TheKMan, TheMightyOrb, TheOldJacobite, Theallpowerfulma, Theeditor93, Thespuddy, Thewikipedian, Thingg, Thorwald, Thylacinus cynocephalus, Tide rolls, Tidying Up, Tigen, TimeClock871, Timneu22, Tinosa, Titoxd, Tk47517, Tnock, Tobby72, Toiyabe, Tomakiv, TommyBoy, TopoChecker, Toxicroak, Toytoy, Tpahl, Tracer9999, Tradnor, Trasel, Travis.Thurston, Tree Biting Conspiracy, Trekphiler, Tresiden, Trimalchio, Triona, Tris189Woot, Truflip99, Tuffluffjimmy, TwinCityIL, Twp, Twsx, Tygar, Tyrol69, Ulmanor, Ulric1313, Umdenken, UncleDouggie, Ųndelope32, Unyoyega, Ursoc, Uyvsdi, Valfontis, Vanished user 5zariu3jisj0j4irj, Vanished user ewfisn2348tui2f8n2fio2utjfeoi210r39jf, Vary, VegaDark, Velella, Velvetron, Versus22, Vesperclint, Vicenarian, Vicki Rosenzweig, VigilancePrime, Virgil61, Viriditas, Vokkare, Vpuliva, Vrenator, Vsmith, Vtskier2010, Vystrix Nexoth, WJBscribe, WVhybrid, Waggers, Wahkeenah, Wallowalake, Walrus heart, Wangi, Wapcaplet, Warpflyght, Wars, Waterjuice, Wavelength, Wayne Slam, Wdflake, Weatherman78, Weregerbil, WestWovles13, WhisperToMe, Wiki alf, Wiki13, Wikiguy008, Wikipelli, Willydick, Winchelsea, Wknight94, Wmahan, Wohuigongfu, Wolfscarr, Wolverineblue, Woohookitty, XRK, Xezbeth, Xxkaenxx, Y5nthon5a, YUL89YYZ, Yamamoto Ichiro, Yamla, YellowMonkey, Yintan, Ylee, Yortek98123, Youngamerican, Yvuogksq, ZX81, ZabMilenko, ZacoTaco, Zafiroblue05, Zainubrazvi, Zaui, Zaxname123, Zeimusu, Zeno293, Zip34532, Zntrip, Zoe, Zonath, Zoopazoop, Zzuuzz, Zzyzx11, A, 1915 anonymous edits

Crook_County,_Oregon *Source*: http://en.wikipedia.org/w/index.php?title=Crook_County%2C_Oregon *Contributors*: 90, AXRL, Accordianate, Ajbenj, Backspace, Bobblewik, Brian H Wilson, Bunchofgrapes, CanisRufus, CommonsDelinker, D6, Eaglecap Backpack, Finetooth, Floydspinky71, Fratrep, Ground Zero, Hyperfast, Junkyardprince, Kaltenmeyer, Leitmotiv, Leslie Mateus, Libereco Li, Lightmouse, Lincolnite, Llywrch, Lotje, Monegasque, Motthoangwehuong, Northwesterner1, Numbo3, Nyttend, Omnedon, Orygun, Owen, Pearle, Perez2022, PurpleSunflower, Ram-Man, Rich Farmbrough, Richwales, Rwentechaney, Sgt Pinback, Smallbones, TUF-KAT, Template namespace initialisation script, Valfontis, Worldenc, 10 anonymous edits

Image Sources, Licenses and Contributors

File:Petefrench.jpg *Source*: http://en.wikipedia.org/w/index.php?title=File:Petefrench.jpg *License*: unknown *Contributors*: Valfontis

Image:Pete French Round Barn interior EncMstr.jpeg *Source*: http://en.wikipedia.org/w/index.php?title=File:Pete_French_Round_Barn_interior_EncMstr.jpeg *License*: unknown *Contributors*: Pete Forsyth and EncMstr

File:SlaterMill.JPG *Source*: http://en.wikipedia.org/w/index.php?title=File:SlaterMill.JPG *License*: unknown *Contributors*: Forest J. Handford

File:LorenAndrus1996.jpg *Source*: http://en.wikipedia.org/w/index.php?title=File:LorenAndrus1996.jpg *License*: unknown *Contributors*: User:Ktacia

File:Stephen Phillips House.jpg *Source*: http://en.wikipedia.org/w/index.php?title=File:Stephen_Phillips_House.jpg *License*: unknown *Contributors*: User:Fletcher6

File:George B. Hartzog Jr.jpg *Source*: http://en.wikipedia.org/w/index.php?title=File:George_B._Hartzog_Jr.jpg *License*: unknown *Contributors*: Auntof6, Choess, IvoShandor

File:Cecil D Andrus.png *Source*: http://en.wikipedia.org/w/index.php?title=File:Cecil_D_Andrus.png *License*: unknown *Contributors*: Bj.schoenmakers, Peteforsyth, Scooter, 1 anonymous edits

Image:Mohonk Mountain House 2011 View of Mohonk Guest Rooms from One Hiking Trail FRD 3205.jpg *Source*: http://en.wikipedia.org/w/index.php?title=File:Mohonk_Mountain_House_2011_View_of_Mohonk_Guest_Rooms_from_One_Hiking_Trail_FRD_3205.jpg *License*: unknown *Contributors*: Fred Hsu on en.wikipedia

File:Robie House.jpg *Source*: http://en.wikipedia.org/w/index.php?title=File:Robie_House.jpg *License*: unknown *Contributors*: User:Rdsmith4

File:HistoricPlacesNationalRegisterPlaque.JPG *Source*: http://en.wikipedia.org/w/index.php?title=File:HistoricPlacesNationalRegisterPlaque.JPG *License*: unknown *Contributors*: User:Jonathunder

File:MLK tomb.JPG *Source*: http://en.wikipedia.org/w/index.php?title=File:MLK_tomb.JPG *License*: unknown *Contributors*: Disavian, Eoghanacht, Gray wolf, Nyttend, Origamiemensch, Skeezix1000, Verica Atrebatum, 1 anonymous edits

File:Flag of Canada.svg *Source*: http://en.wikipedia.org/w/index.php?title=File:Flag_of_Canada.svg *License*: unknown *Contributors*: Anomie

File:Flag of Germany.svg *Source*: http://en.wikipedia.org/w/index.php?title=File:Flag_of_Germany.svg *License*: unknown *Contributors*: Anomie

File:Flag of Hong Kong.svg *Source*: http://en.wikipedia.org/w/index.php?title=File:Flag_of_Hong_Kong.svg *License*: unknown *Contributors*: Designed by

File:Flag of Italy.svg *Source*: http://en.wikipedia.org/w/index.php?title=File:Flag_of_Italy.svg *License*: unknown *Contributors*: Anomie

File:Flag of the Netherlands.svg *Source*: http://en.wikipedia.org/w/index.php?title=File:Flag_of_the_Netherlands.svg *License*: unknown *Contributors*: User:Zscout370

File:Flag of New Zealand.svg *Source*: http://en.wikipedia.org/w/index.php?title=File:Flag_of_New_Zealand.svg *License*: unknown *Contributors*: Achim1999, Adabow, Adambro, Arria Belli, Avenue, Bawolff, Bjankuloski06en, ButterStick, Cycn, Denelson83, Donk, Duduziq, EugeneZelenko, Fred J, Fry1989, Hugh Jass, Ibagli, Jusjih, Klemen Kocjancic, Mamndassan, Mattes, Nightstallion, O, Peeperman, Poromiami, Reisio, Rfc1394, Sarang, Shizhao, Tabasco, Transparent Blue, Väsk, Xufanc, Zscout370, 37 anonymous edits

File:Flag of Portugal.svg *Source*: http://en.wikipedia.org/w/index.php?title=File:Flag_of_Portugal.svg *License*: unknown *Contributors*: Columbano Bordalo Pinheiro (1910; generic design); Vítor Luís Rodrigues; António Martins-Tuválkin (2004; this specific vector set: see sources)

File:Flag of Romania.svg *Source*: http://en.wikipedia.org/w/index.php?title=File:Flag_of_Romania.svg *License*: unknown *Contributors*: User:AdiJapan

Image:Round barn, Hancock Shaker village.jpg *Source*: http://en.wikipedia.org/w/index.php?title=File:Round_barn,_Hancock_Shaker_village.jpg *License*: unknown *Contributors*: User:Deadstar, User:Eranb

File:Fountaingrove Round Barn.jpg *Source*: http://en.wikipedia.org/w/index.php?title=File:Fountaingrove_Round_Barn.jpg *License*: unknown *Contributors*: User:Frank Schulenburg

Image:Octabarn.jpg *Source*: http://en.wikipedia.org/w/index.php?title=File:Octabarn.jpg *License*: unknown *Contributors*: Original uploader was Motorrad-67 at en.wikipedia Later versions were uploaded by Liftarn at en.wikipedia.

File:RoundBarn.jpg *Source*: http://en.wikipedia.org/w/index.php?title=File:RoundBarn.jpg *License*: unknown *Contributors*: User:Jonathunder

Image:TenSidedBarn.jpg *Source*: http://en.wikipedia.org/w/index.php?title=File:TenSidedBarn.jpg *License*: unknown *Contributors*: User Royalbroil on en.wikipedia

File:Harney_County_Oregon_Incorporated_and_Unincorporated_areas_Burns_Highlighted.svg *Source*: http://en.wikipedia.org/w/index.php?title=File:Harney_County_Oregon_Incorporated_and_Unincorporated_areas_Burns_Highlighted.svg *License*: unknown *Contributors*: Arkyan

Image:BroadwayBurnsOR.jpg *Source*: http://en.wikipedia.org/w/index.php?title=File:BroadwayBurnsOR.jpg *License*: unknown *Contributors*: User:Cacophony

Image:ORMap-doton-Burns.png *Source*: http://en.wikipedia.org/w/index.php?title=File:ORMap-doton-Burns.png *License*: unknown *Contributors*: Original uploader was Bumm13 at en.wikipedia

Image:Harney Basin near Burns, Oregon, 2007.jpg *Source*: http://en.wikipedia.org/w/index.php?title=File:Harney_Basin_near_Burns,_Oregon,_2007.jpg *License*: unknown *Contributors*: tetraconz

File:HarneyCountyCourthouse.jpg *Source*: http://en.wikipedia.org/w/index.php?title=File:HarneyCountyCourthouse.jpg *License*: unknown *Contributors*: User:Cacophony

File:Map of Oregon highlighting Harney County.svg *Source*: http://en.wikipedia.org/w/index.php?title=File:Map_of_Oregon_highlighting_Harney_County.svg *License*: unknown *Contributors*: User:Dbenbenn

File:Harney.gif *Source*: http://en.wikipedia.org/w/index.php?title=File:Harney.gif *License*: unknown *Contributors*: Eaglecap Backpack

File:Harney County, Oregon, sign.JPG *Source*: http://en.wikipedia.org/w/index.php?title=File:Harney_County,_Oregon,_sign.JPG *License*: unknown *Contributors*: Ken Lund

File:Lake County, Oregon courthouse.jpg *Source*: http://en.wikipedia.org/w/index.php?title=File:Lake_County,_Oregon_courthouse.jpg *License*: unknown *Contributors*: Ken Lund

File:Lake.gif *Source*: http://en.wikipedia.org/w/index.php?title=File:Lake.gif *License*: unknown *Contributors*: Eaglecap Backpack

Image:AbertRim-right.jpg *Source*: http://en.wikipedia.org/w/index.php?title=File:AbertRim-right.jpg *License*: unknown *Contributors*: Jsayre64

Image:Oregon Population Growth by County.png *Source*: http://en.wikipedia.org/w/index.php?title=File:Oregon_Population_Growth_by_County.png *License*: unknown *Contributors*: User:Northwest-historian

Image:The Three Sisters from Peter Skene Ogden State Park.jpg *Source*: http://en.wikipedia.org/w/index.php?title=File:The_Three_Sisters_from_Peter_Skene_Ogden_State_Park.jpg *License*: unknown *Contributors*: User:Stratosphere

Image:DSCN7240 roadsouthofbend e 300.jpg *Source*: http://en.wikipedia.org/w/index.php?title=File:DSCN7240_roadsouthofbend_e_300.jpg *License*: unknown *Contributors*: User:Decumanus

File:Seal of Oregon.svg *Source*: http://en.wikipedia.org/w/index.php?title=File:Seal_of_Oregon.svg *License*: unknown *Contributors*: User:Svgalbertian

File:Map of Oregon NA.png *Source*: http://en.wikipedia.org/w/index.php?title=File:Map_of_Oregon_NA.png *License*: unknown *Contributors*: Ciaurlec, Huebi, Jmabel, Look2See1, Peteforsyth

File:OregonCoastEcola.jpg *Source*: http://en.wikipedia.org/w/index.php?title=File:OregonCoastEcola.jpg *License*: unknown *Contributors*: User:Cacophony

file:speakerlink-new.svg *Source*: http://en.wikipedia.org/w/index.php?title=File:Speakerlink-new.svg *License*: unknown *Contributors*: User:Kelvinsong

File:Panorama of Portland, Oregon, in 1890, 1890 - NARA - 530891.jpg *Source*: http://en.wikipedia.org/w/index.php?title=File:Panorama_of_Portland,_Oregon,_in_1890,_1890_-_NARA_-_530891.jpg *License*: unknown *Contributors*:

File:Trilliumlake.jpg *Source*: http://en.wikipedia.org/w/index.php?title=File:Trilliumlake.jpg *License*: unknown *Contributors*: Notwist, Pauk, Shyam, Ö

File:Crater lake oregon.jpg *Source*: http://en.wikipedia.org/w/index.php?title=File:Crater_lake_oregon.jpg *License*: unknown *Contributors*: User:Zainubrazvi

File:Sunset over Malheur Butte near Vale, Oregon.JPG *Source*: http://en.wikipedia.org/w/index.php?title=File:Sunset_over_Malheur_Butte_near_Vale,_Oregon.JPG *License*: unknown *Contributors*: Staplegunther (talk)David Jolley. Original uploader was Staplegunther at en.wikipedia

File:Eugene skyline crop.jpg *Source*: http://en.wikipedia.org/w/index.php?title=File:Eugene_skyline_crop.jpg *License*: unknown *Contributors*: User:Jsayre64

File:Public-Lands-Western-US.png *Source*: http://en.wikipedia.org/w/index.php?title=File:Public-Lands-Western-US.png *License*: unknown *Contributors*: User:Northwest-historian

File:Wpdms_shdrlfi020l_willamette_valley.jpg *Source*: http://en.wikipedia.org/w/index.php?title=File:Wpdms_shdrlfi020l_willamette_valley.jpg *License*: unknown *Contributors*: Mattes, Schimmelreiter, TommyBee

File:Oregon State Capitol 1.jpg *Source*: http://en.wikipedia.org/w/index.php?title=File:Oregon_State_Capitol_1.jpg *License*: unknown *Contributors*: User:Aboutmovies

File:Oregon voter reg 1950-2006.png *Source*: http://en.wikipedia.org/w/index.php?title=File:Oregon_voter_reg_1950-2006.png *License*: unknown *Contributors*: Peteforsyth

File:Grass seed warehouse.jpg *Source*: http://en.wikipedia.org/w/index.php?title=File:Grass_seed_warehouse.jpg *License*: unknown *Contributors*: User:Owen

File:Craft Breweries Per Capita (US).svg *Source*: http://en.wikipedia.org/w/index.php?title=File:Craft_Breweries_Per_Capita_(US).svg *License*: unknown *Contributors*: User:Alataristarion, User:EdwardRech, User:Northwest-historian

File:Oregon county pop 2007.png *Source*: http://en.wikipedia.org/w/index.php?title=File:Oregon_county_pop_2007.png *License*: unknown *Contributors*: User:Dbenbenn

File:Oregon Population Growth by County.png *Source*: http://en.wikipedia.org/w/index.php?title=File:Oregon_Population_Growth_by_County.png *License*: unknown *Contributors*: User:Northwest-historian

File:Memorial Union at Oregon State University.jpg *Source*: http://en.wikipedia.org/w/index.php?title=File:Memorial_Union_at_Oregon_State_University.jpg *License*: unknown *Contributors*: Jareha, Owen, Werewombat

File:RoseGardenArenaInterior3.jpg *Source*: http://en.wikipedia.org/w/index.php?title=File:RoseGardenArenaInterior3.jpg *License*: unknown *Contributors*: User:Cacophony

File:Flag of the People's Republic of China.svg *Source*: http://en.wikipedia.org/w/index.php?title=File:Flag_of_the_People's_Republic_of_China.svg *License*: unknown *Contributors*: User:Denelson83, User:SKopp, User:Shizhao, User:Zscout370

File:Flag of Japan.svg *Source*: http://en.wikipedia.org/w/index.php?title=File:Flag_of_Japan.svg *License*: unknown *Contributors*: Anomie

File:Flag of South Korea.svg *Source*: http://en.wikipedia.org/w/index.php?title=File:Flag_of_South_Korea.svg *License*: unknown *Contributors*: Various

File:Flag of Iraq.svg *Source*: http://en.wikipedia.org/w/index.php?title=File:Flag_of_Iraq.svg *License*: unknown *Contributors*: User:Hoshie, User:Militaryace

File:CrookCountyCourthouse.jpg *Source*: http://en.wikipedia.org/w/index.php?title=File:CrookCountyCourthouse.jpg *License*: unknown *Contributors*: User:Cacophony

File:Map of Oregon highlighting Crook County.svg *Source*: http://en.wikipedia.org/w/index.php?title=File:Map_of_Oregon_highlighting_Crook_County.svg *License*: unknown *Contributors*: User:Dbenbenn

MIX
Papier aus verantwortungsvollen Quellen
Paper from responsible sources
FSC® C105338

Printed by Books on Demand GmbH, Norderstedt / Germany